21st CENTURY SCIENCE & TECHNOLOGY

Feb. 2014 Special Report // Physical Chemistry: The Continuing Gifts of Prometheus

Contents

3 — OVERVIEW // **The Gifts of Prometheus: Physical Chemistry and Nuclear Fusion**
Jason Ross

6 — **Measuring Fire: Energy Flux Density**
Benjamin Deniston

9 — PHYSICAL CHEMISTRY // A HISTORY

10 — **Metallurgy: The Birth of Physical Chemistry**
Jason Ross

19 — **Chemistry: The Active Power of the Elements**
Michael Kirsch

25 — **Electromagnetism: A New Dimension**
Creighton Cody Jones

32 — **The Nuclear Era: Man Controls the Atom**
Liona Fan-Chiang

39 — PHYSICAL CHEMISTRY // THE PROMETHEAN FUTURE

40 — **NAWAPA and Continental Water Management: A Promethean Task**
Jason Ross

44 — **Helium-3: Stealing the Sun's Fire**
Natalie Lovegren

54 — APPENDIX // **Prometheus: The Historical Record**
Jason Ross

EDITORIAL STAFF

Editor-in-Chief
Jason Ross

Managing Editor
Marsha Freeman

Associate Editor
Christine Craig

Staff Writers
Megan Beets
Benjamin Deniston
Liona Fan-Chiang
Creighton Jones
Michael Kirsch
Natalie Lovegren
Meghan Rouillard

21st Century Science & Technology (ISSN 0895-6820) is an occasional publication by 21st Century Science Associates, 60 Sycolin Road, Suite 203, Leesburg, VA 20175.
Tel. (703) 777-6943.

Address all correspondence to:
21st Century, P.O. Box 16285, Washington, D.C. 20041.

Letters to the editor:
letters@21stcenturysciencetech.com

21st Century is dedicated to the promotion of unending scientific progress, all directed to serve the proper common aims of mankind.

Opinions expressed in articles are not necessarily those of 21st Century Science Associates.

We are not responsible for unsolicited manuscripts.

www.21stcenturysciencetech.com

© 2014 21st Century Science Associates

The Gifts of Prometheus
Physical Chemistry and Nuclear Fusion

PROMETHEUS: But of wretched mortals he [Zeus] took no notice, desiring to bring the whole race to an end and create a new one in its place. Against this purpose none dared make stand except me—I only had the courage; I saved mortals so that they did not descend, blasted utterly, to the house of Hades. This is why I am bent by such grievous tortures, painful to suffer, piteous to behold.

CHORUS: Did you transgress even somewhat beyond this offense?

PROMETHEUS: What's more, I gave mankind fire.

CHORUS: What! Do mortals now have flame-eyed fire?

PROMETHEUS: Yes, and from it they shall learn many arts.

— Aeschylus, *Prometheus Bound*

The Special Report you are now reading, "Physical Chemistry: The Continuing Gifts of Prometheus" serves two purposes.

The first, is to impart a living, joyous sense of the difference between mere money and true value. The qualitative nature of real human advancement is best seen in broad terms by looking at the changing use of fire, from which Prometheus says man "shall learn many arts."

From wood to coal to nuclear power, the *platforms* for activity provided by these power sources mark successive stages of human economic development. In this report, we will use the development of physical chemistry, whose origins stretch to the beginning of human prehistory, with the uses of fire to change materials, from the birth of metallurgy to today's semiconductors and nuclear science, to give an image of true physical value.

The second purpose, is to sketch out the foundation for a human future based upon this concept of Promethean value. Value, which lies in what will be brought about in the future, can always be expressed in specific, wide-ranging goals. The specific goals that will measure the depth of our powers to develop will be covered briefly: the development of controlled nuclear fusion and the implementation of continental water management. Reference is made to our previous Special Report: "Nuclear NAWAPA XXI: Gateway to the Fusion Economy."[1]

Prometheus was a true non-mythical historical personality, who endured the wrath of the god Zeus for daring to bring "fire" from heaven to man (along with poetry, astronomy, and science in general). Though chained by Zeus to a rock to have the torture of an eagle devouring his liver every day, Prometheus was unawed by Zeus's power to punish him, and held him in utter contempt. The story of the Olympian god Zeus and Prometheus the Fire-Bringer is not fictional, not a piece of idle drama. Here we find the most pure expression of the fight that has dominated large-scale political and economic conflict throughout mankind's existence. We find the essence of the confrontation between an oligarchical outlook, in which some few rulers maintain capricious power over (preferably stupefied) masses, and the humanist outlook—in which the true identity of every human being as a potential genius is embraced and in which providing the opportunity (physical, moral, and emotional) for every individual to lead a functionally *immortal* life is the ultimate goal.

"Every art possessed by man comes from Prometheus."[2]

Our exploration of the successful applications of this Promethean power will take us through four main fields, which can all be grouped under the general concept of *physical chemistry*. These fields are: metallurgy, the birth of modern chemistry, the world of electromagnetism, and the science of the nucleus. After our voyage, we'll be able to reach new conclusions.

Metallurgy, the great science of transforming rocky ores into useful metals, came into being roughly at the birth of history, of man's written records of his doings. While some few metals, such as gold and occasionally copper exist in pure states, the birth of the Bronze Age marked the advent of extractive metallurgy: transforming the green stone malachite into shining copper, and producing an entirely new metal, bronze, by combining copper with tin. From bronze to iron to steel, the transformation of ores into increasingly specialized metal alloys requires tremendous amounts of heat and the transportation logistics to move ores, flux, and fuel to processing sites.

As modern science was born, with the work of Nicolaus of Cusa and his follower Johannes Kepler, the alchemy of the Middle Ages (which sought to "make money"—gold) was replaced by the science of chemistry, which sought to understand the powers and activities of *all* physical substances, to increase man's power over them. Specific properties, sometimes found in only a few substances or elements, were found to represent general principles of nature. Understanding the elements of nature by their *potentials to act*, rather than their observable properties was key. Antoine Lavoisier's demonstration that heat was not a substance, the periodic system of the elements (based on the comparison of their potentials for action and atomics weights), developed by Dmitri Mendeleev, and the use of electricity to pry apart previously inseparable elements are a few of the examples that illustrate man's increasing knowledge of, and power over, the principles that govern nature in the small. This *power* is valuable; it is true economic wealth.

Man's use of electricity and the electromagnetic properties of materials has developed dramatically: from the knowledge that rubbing amber (*elektron* in Greek) could cause it to attract small bits of lint, to the development of the first machines capable of producing and storing "electric fluid" as it was then known, to the use of chemical reactions (batteries) to produce *running* ("current") electric fluid, to the generator and the motor, and to the use of semiconductors in today's solid-state computing equipment. Who could have imagined that the attractive power of rubbed amber would be used as the primary source of industrial motion (via the motor), or form the basis for automated control systems for industry via the integrated circuit?

Nuclear science, which arose from humble beginnings of uranium salts and photographic plates, brought to light properties of matter that lay below the scale dealt with by chemistry. It offers the greatest as-yet-unrealized potential to transform human life through remarkable properties of matter that are invisible to the eyes of chemistry and electricity. We see this in the power inherent in the ability of the uranium nucleus to fission, the new materials based on isotopic specificity, and the enormous potentials for fusion power. Fusion will fundamentally transform our relationship to materials in a way unmatched since the original birth of metallurgy, and can provide the power basis required to develop a system for planetary defense from errant asteroids and comets.

---

1. See http://21stcenturysciencetech.com/Nuclear_NAWAPA.html

2. Heading quotes are Prometheus speaking in Aeschylus's play.

"I caused unseen hopes to dwell within their breasts."

The full implementation of these myriad gifts of Prometheus *currently allows the very rapid elimination of poverty, worldwide.* Why, then, have these gifts been withheld? Why have they not been put to use? Surely, it is not by accident. The same species that has developed these powers certainly has the ability to put them fully into practice. Is there a physical law that has prevented the use of fertilizer, irrigation techniques, and modern harvesting technology in poor areas of the world? Is there a law of nature that prevents the application of nuclear fission power and the development of fusion? Would a Promethean society spend more on movies, video games, and gambling than on the scientific breakthroughs that will define human history for generations to come?[3] Is it *natural* to prefer "traditional" beliefs to new discoveries? Is it *natural* to orient to base pleasures, rather than those of the mind? Is it *natural* for humanity to act as though it were another species, devoid of reason?

The *unnatural* cause of these uncharacteristic behaviors is the presence of what may be considered another species: the oligarchical species of man. This pestilential social ill, the presence of an oligarchy, and its too-general toleration, has held mankind back for millenia through empires, wars, the suppression of science and culture, and myths of overpopulation dating back literally *thousands* of years.

It is commonly said that the Dark Ages followed the collapse of the Roman Empire. In reality, the Roman Empire, which existed by looting and slavery, rather than technological advancement, *was* a dark age. The Byzantine Empire developed ornate palaces, but no technological improvement. It was in the early part of the second millenium, a lull between empires, before the full establishment of Venice as an imperial power, that the development of the great cathedrals of Europe began, and the first structure taller than the great pyramid of Giza was built.

The formation of the United States represented the aspirations of those in Europe seeking the means to rid themselves of oligarchism, and develop society according to the pursuit of happiness: bringing about the increasing perfection of others. The origins of this project, initiated by circles around Nicholas of Cusa, saw substantial, but ultimately, temporary, victories in the Revolutionary War, the Constitution, the policies of Alexander Hamilton, John Quincy Adams, Abraham Lincoln, and other more recent leaders. But the most noble aspirations of those who formed this nation will not have been achieved so long as oligarchism exists on this planet—so long as Zeus has not been defeated.

"Though they had eyes to see, they saw to no avail."

What really matters? What matters to us of people from three millenia ago? Those who developed bronze or made their lives possible contributed something of unquestionably durable importance to human civilization, an *evolution* of the species: not a genetic evolution, but a super-genetic one. What do the lives of those who wasted their potential in dissipating pleasures mean to us today? What opportunity for long-lasting contributions are afforded to those subject to grinding poverty, unable, by their conditions of life, to develop their mental faculties?

Truly, creating the conditions for the elevation of all members of the human race, to being meaningfully *human*, is the greatest of political goals, and the most noble aspiration for the life of any individual. This is the Promethean outlook, and it can no longer coexist with the oligarchical.

— Jason Ross

---

3. Americans now spend $10 billion annually just at movie theaters, and more than that on electronic games. NASA's budget is currently $18 billion. Meanwhile, Americans lost $119 billion gambling in 2013.

Video: *The Gifts of Prometheus*

Creighton Jones — Jason Ross — Liona Fan-Chiang

View the Feb. 1, 2014 video webcast at: **larouchepac.com/GiftsOfPrometheus-Webcast**

Measuring Fire:
Energy Flux Density

by Benjamin Deniston

We begin with the first of the gifts of Prometheus, fire, from which he says man "shall learn many arts." The earliest archaeological distinction between mankind and the apes comes with the first appearance of ancient fire pits, used to control the power of fire for the betterment of the conditions of life of those wielding that new power.

From that time onward, mankind could no longer be characterized biologically or by biological evolution—the evolution of the creative mental powers unique to the human mind became the determining factor. Biology took a backseat to the increased power of thought wielded by the human species.

This is the secret—and science—of economic growth, expressed through the control over successively higher forms of fire. This started with transitions to more energy-dense forms of chemical fire, from simple wood burning, to charcoal, then to coal and coke, and onto petroleum and natural gas. Each of these new types depended upon new chemical reactions, which not only provided the potential for a more energy dense form of fire, but opened up *new domains of control and utilization of matter*. Metallurgy, materials development, and physical chemistry all developed in dynamic interaction with the development of new forms of fire.

The revolutionary discoveries around the turn of the 20th century showed mankind an immense potential entirely beyond chemical reactions: the fundamental equivalence of matter and energy, as expressed in the domains of fission, fusion, and matter-antimatter reactions. Each in this series of relativistic reactions (reflecting the relationship betweeen mass and energy developed by Einstein) operates at successively higher energy densities—and the entire set is orders of magnitude beyond the entire set of successive chemical reactions.[1] While this distinction is usefully expressed in the immense difference in the quantity of energy released in nuclear versus chemical reactions, the measured quantitative difference is the effect of a qualitatively distinct, higher domain of action.

Control over higher energy densities enables the increase in what Lyndon LaRouche has identified as the energy flux density of the economy, as can be measured by the rate of energy use per person and per unit area of the economy as a whole. This increasing power is associated with qualitative changes throughout the entire society—fundamentally new technologies, new resource bases, new levels of living standards, and, what are fundamentally new economies.

Table I: The Energy Density of Fuels	
FUEL SOURCE	**ENERGY DENSITY (J/g)**
Combustion of Wood	1.8×10^4
Combustion of Coal (Bituminous)	2.7×10^4
Combustion of Petroleum (Diesel)	4.6×10^4
Combustion of H_2/O_2	1.3×10^4 (full mass considered)
Combustion of H_2/O_2	1.2×10^5 (only H_2 mass considered)
Typical Nuclear Fuel	3.7×10^9
Direct Fission Energy of U-235	8.2×10^{10}
Deuterium-Tritium Fusion	3.2×10^{11}
Annihilation of Antimatter	9.0×10^{13}

Fuel energy densities. The change from wood to matter-antimatter reactions is so great that progress must be counted in orders of magnitude, and the greatest single leap is seen in the transition from chemical to nuclear processes.

---

1. This is why individual nuclear explosives, even small ones, are measured in terms of thousands of tons, or even *millions of tons* of TNT. The largest thermonuclear weapon ever detonated, the Soviet Union's 1961 Tsar Bomba, was a 50 megaton explosion, meaning it would take the explosion of 50 million tons of TNT to release that much energy from chemical reactions. The Tsar Bomba was a single bomb, dropped from a single airplane (tested over an unpopulated region far north), while 50 million tons of TNT would fill 100 oil supertankers.

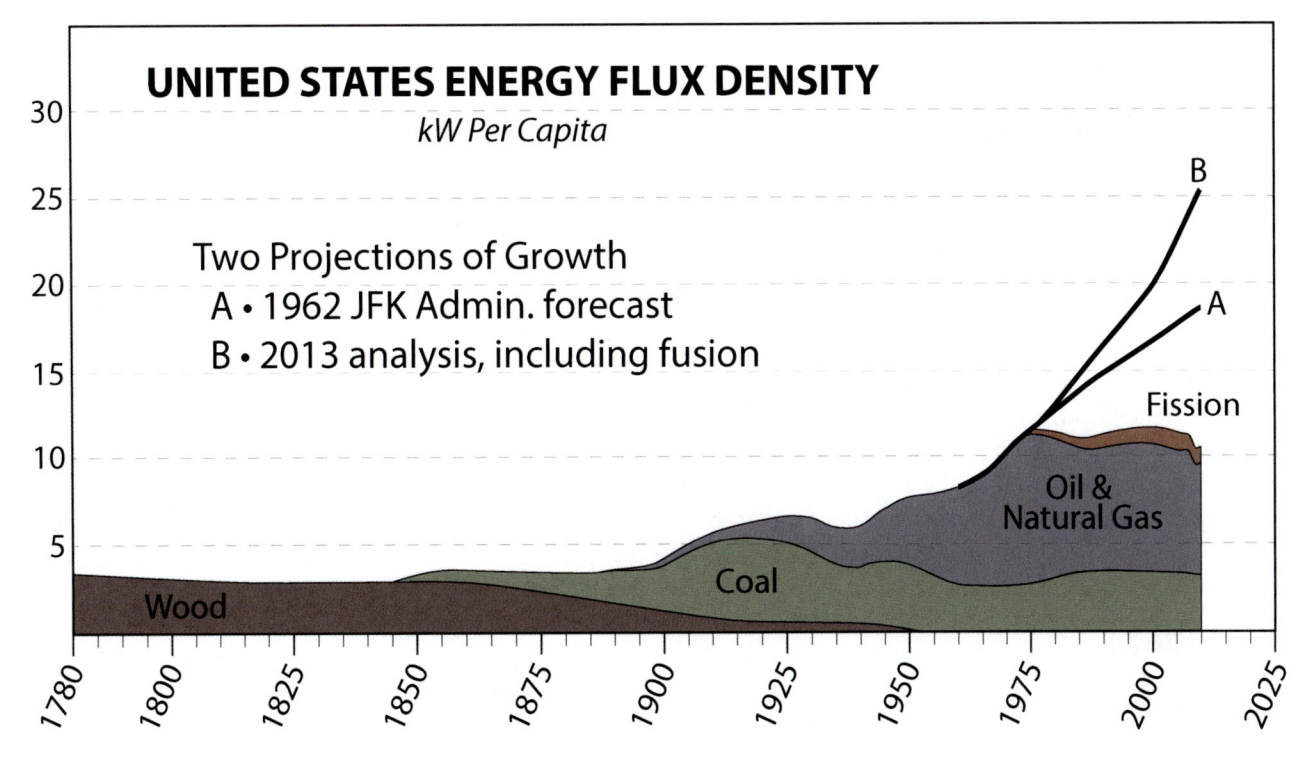

UNITED STATES ENERGY FLUX DENSITY
kW Per Capita

Two Projections of Growth
A • 1962 JFK Admin. forecast
B • 2013 analysis, including fusion

Graphic by Benjamin Deniston, data from U.S. Energy Information Administration and from "Civilian Nuclear Power, a Report to the President" submitted to JFK by Leland Haworth.

Per capita power consumption for the United States from 1780 to 2010, divided by the major sources of power. The general growth trend is clear, until 1970, when the zero-growth insanity took over the United States. Two projections indicate what could and should have happened. Curve A is a 1962 projection made by the John F. Kennedy administration, which focused on the then-coming role of nuclear fission power. Curve B is an estimation of what was possible if the Kennedy vision had been pursued, followed by the development of controlled thermonuclear fusion (following the 1970s realization of the feasibility of fusion). These two curves, compared with the actual levels, show the 40-year growth gap which is a major source of the current economic collapse.

A Short History of Energy Flux Density

Start with the simple rate of biological energy usage for the human body, which is, very roughly, 100 watts (corresponding to consuming 2,000 food calories a day). Assuming a hypothetical pre-fire civilization in which all work is performed by human muscle, the power employed to sustain the "economy"—the power of labor—is 100 watts per capita.

Compare this with the growing per capita power usage throughout the history of the United States.

At the time of the nation's founding, the wood-based economy provided around 3,000 watts per capita. In this wood-based economy, the effective power that each individual wielded and represented, through the active use and application of the heat provided by the burning of wood and charcoal, was thirty times higher than the simple muscle power of a hypothetical fire-less society. This was not just "more" energy, but a quality of energy that enabled people to create new states of matter and chemistry, states which could never be created by muscle power alone.[2]

The increasing use of coal throughout the economy raised the power to over 5,000 watts per capita by the 1920s. Each individual then expressed nearly twice the power of the wood-based economy, supporting the heat-powered machinery and transportation which revolutionized the economy, and the development of modern chemistry enabled the beginnings of the greatest revolution in mankind's understanding of and control over matter since the actions of Prometheus.

By 1970, the use of petroleum and natural gas had brought power to over 10,000 watts per capita—100 times the per capita power of our hypothetical fire-less society. With each transition, the previous fuel base declined as a power source, allowing it to be used for things other than combustion, as wood is used for construction, and petroleum should be reserved for plastics and related noncombustible products of the petrochemical industry.

---

2. As is exceptionally clear in the history of metallurgy, for example. No amount of muscle power can convert ore into metal.

Zeus Today

Nuclear fission power was fully capable of sustaining and accelerating the U.S. historical growth rate well into the 21st century. In a conservative estimate, based upon previous growth rates and the potential of nuclear power, fission should have brought the U.S. economy to the range of 20,000 watts per capita by some time before the year 2000.[3]

By then, assuming the nation had maintained a pro-growth orientation, as fission power was becoming the dominant power source, the beginnings of applied fusion power should have begun to emerge. With ocean water as a source for an effectively limitless fuel source for fusion reactors (deuterium), the U.S. economy would have been on a path to an energy flux density of around 40,000 watts per capita, and beyond, in the first generation of the 21st century, four times the current value of 10,000 watts. Virtually every single concern over resource limitations (from food, to water, to metals, etc.) and energy limitations for all mankind, across the entire planet, would be solved with a fusion economy—and that for many generations to come.

However, this natural growth process was intentionally stopped by the resurgence of Zeus, in the form of the anti-progress, zero-growth environmentalist movement. Imposed on the United States by the exact same Anglo-Dutch empire against which Franklin Roosevelt fought,[4] this green policy has sent the economy on the direct path into the attritional collapse being experienced now—a collapse process accelerated by policies which lower the energy flux density of the economy.

As is clear in the graph, nuclear fission power was never allowed to realize its full potential, and the energy flux density of the economy stagnated, and began to collapse.

The 40-year gap between the needed growth rate and present levels expresses the source of the current economic breakdown, and demonstrates the immediate need for a crash program to develop and implement the next stage, the fusion economy, to overcome decades of lost time by creating a new economy at a higher level than ever before.

Increasing qualities of power—of "fire"—is the essential characteristic of mankind. Either mankind continues to progress, expanding to new levels and higher platforms, or mankind will cease to exist, as Zeus demanded. This is the key to the future, and the past history of mankind.

We now treat four dimensions of physical chemistry: the physical work of metallurgy, chemical characteristics of the elements, electromagnetism, and the nuclear world, which is itself key to our future development of the great Promethean gift upon which the future existence of all mankind absolutely depends—fusion.

---

3. If a serious economic policy had governed the nation following World War II (as was intended by Franklin Roosevelt, but reversed by the presidency of Harry Truman), a higher level could have been reached faster.

4. For a brief overview of the continuity of imperial-genocidal policy of the Ango-Dutch empire, from before the American Revolution to the present day, see "Behind London's War Drive: A Policy To Kill Billions," *EIR*, November 18, 2011.

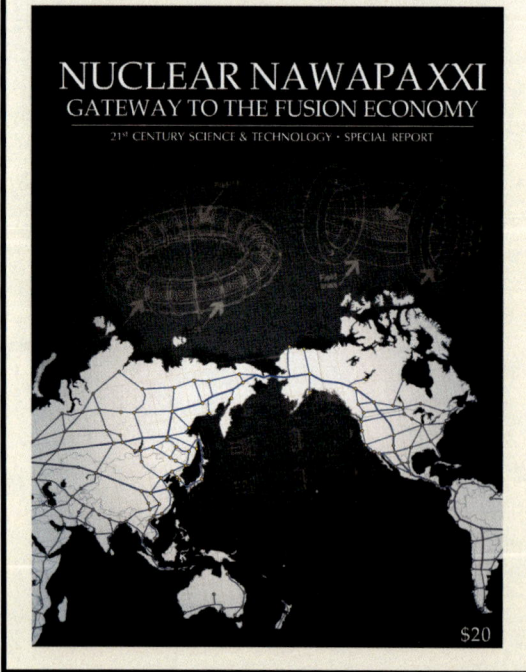

Participate in creating the coming fusion economy. Get a copy of the special report "Nuclear NAWAPA XXI: Gateway to the Fusion Economy," available online:

www.21stcenturysciencetech.com

Print copies available for $20, with free U.S. shipping.

Part 1

Physical Chemistry
A History

Metallurgy
The Birth of Physical Chemistry

Chemistry
The Active Power of the Elements

Electromagnetism
A New Dimension

The Nuclear Era
Man Controls the Atom

Metallurgy
The Birth of Physical Chemistry

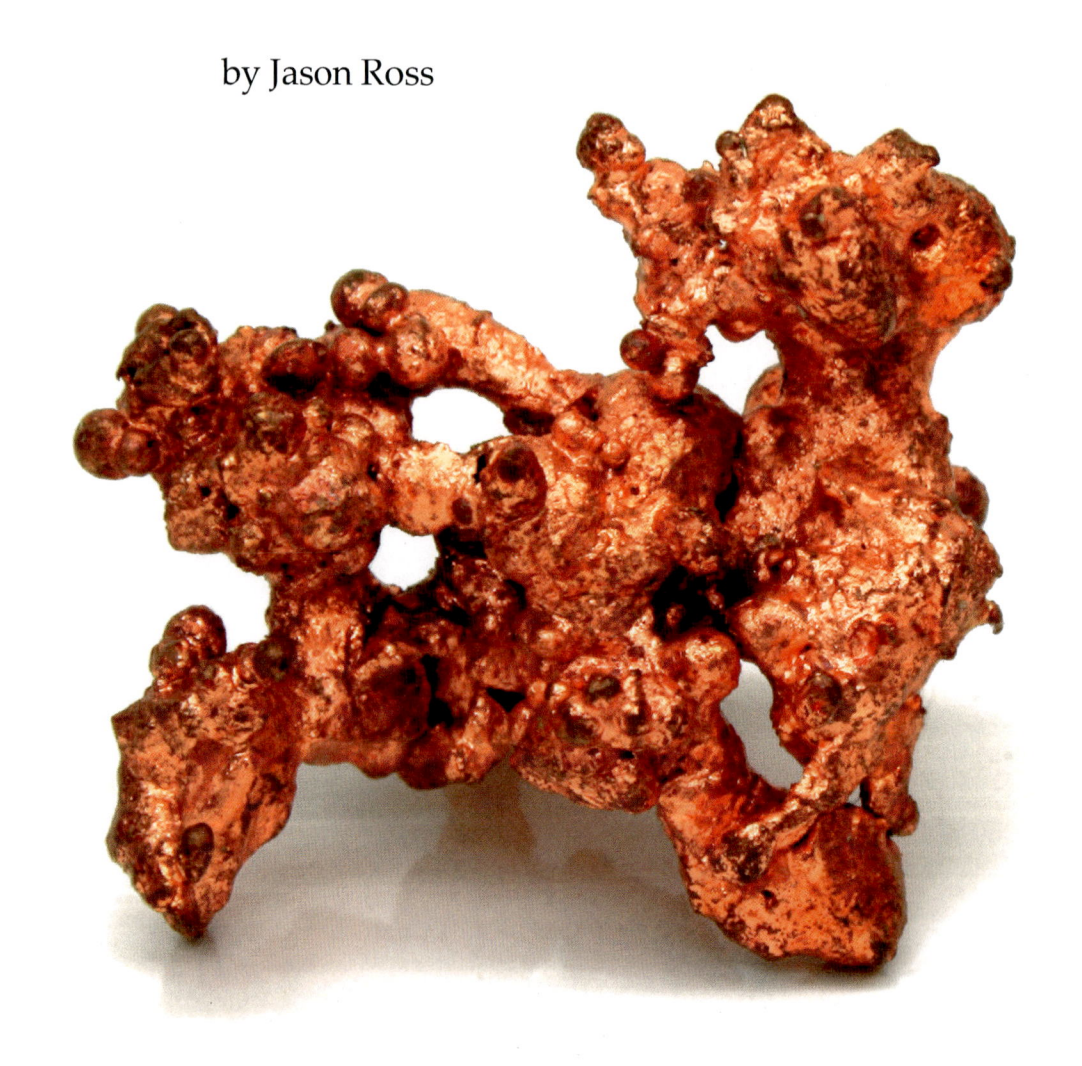

by Jason Ross

A sample of native copper. Some metals, such as copper, gold, and silver, are found natively, in small quantities, in their metallic form on Earth.

Modern civilization makes extensive use of metals for tool-making, structural, and electronic purposes, yet the origins of the bold power of the human mind to create lustrous metal from dull stone are almost completely unknown to most people. We use steel in automobiles and the frames of buildings, nails for carpentry, wires for electricity, pipes for water, metal cans and aluminum foil for food, rivets and zippers in clothing, and jewelry. The casual disposal of aluminum foil after one use would amaze any chemist from the 1800s, when it was one of the most difficult metals to produce.[1]

The development of metallurgy required many individual techniques, from trade in individual metals and ores to water pumps for mines, from prospecting to smithing, but, above all, it required the application of absolutely tremendous amounts of heat. While a wood fire burns hot enough to cook meat (and kill the parasites within it), the temperature is not sufficient to melt copper and produce bronze for casting. For this, the higher energy density of charcoal is required. Every town had its charcoal makers, who would produce the fuel by partially burning wood in an oxygen-poor environment—a pile of smoldering wood covered with turf. The resulting charcoal burned much hotter, and much cleaner than did the original wood. The heats achievable with charcoal fires allowed the working of bronze, and, with the centuries later technique of blast furnaces, which forced more air into the fire, the heats required to melt even iron and steel.

Modern steel production makes use of precise chemical assays to control the processes of alloying and managing carbon content, allowing for specialty steels with unique properties for different environments, such as stainless steel, steel meant to be used underwater, and ultra high-quality steel for such applications as aerospace.[2] Totally new techniques for metallurgy, such as plasma processing with magnetic separation of metal from oxygen, could dramatically reduce the complexity of the process, making in-situ resource utilization in space a real possibility.

The development of metallurgy, from pre-history to today to the future, provides a thrilling image of man the creator, and one of our greatest uses of "fire." Without the power required for processing buried ores into specialty alloys, we'd literally be back in the stone age!

"Native" Metals

In the so-called Stone Age (human history up to approximately 3200 BC in Europe), fire was used for cooking, baking, wood-working, pottery, hardening of stone tools, land-clearing, heat, and light. The use of fire was extended to working with those metals known to the ancients. Even before the advent of extractive metallurgy around 3200 BC, there were certain kinds of metals which could be found in a pure, "native" state. These included gold, copper, silver, and, in the form of meteorites, even iron.[3] Unlike other materials, gold was lustrous, did not decay or corrode, and could be shaped into any form desired by hammering. Copper tools could be made nearly as sharp as stone tools, but could last longer. Incredibly rare meteoritic iron was used for daggers and ornamentation in ancient Egypt, and although inconceivable today, copper was found in nature, as one might find a quartz rock in our day.

These first metals led to the development of the first

left: Hannes Grobe

Silvery-gray galena (left), has an appearance that is both metallic and crystalline, giving an indication of the metal (lead) it contains. In contrast, the azurite (blue) and malachite (green) give no indication of the copper that can be produced from them.

---

1. While today a common metal, aluminum was so difficult to produce without modern methods of electrolysis, that it was the then-exotic and valuable metal used to cap the Washington Monument, and used by Napoleon III for his most honored guests, while others had to eat off of mere gold!

2. Take, as an example, the hooks on aircraft carrier-launched planes. A pilot landing on an aircraft carrier uses the plane's tailhook to hook onto one of the series of "arresting wires" that are fastened across the deck. These arresting wires are made of high-tensile steel that can stop a 54,000-pound aircraft traveling at 150 miles per hour in only two seconds.

3. The Egyptians called weapons formed from meteoritic iron "daggers from heaven."

metal-working skills: hammering, curling, and the use of fire to anneal metal which had become hard by hammering.[4] In search of more of these metals, mines were created, in which pure veins of valuable materials such as gold could be gathered.[5]

Yet, most of the metals used today do *not* come from pure veins: they do not come from native metals. Rather, they are created from *ores*. But most ores don't look the least bit metallic. While it may be no surprise that metallic galena (PbS) was a source for lead, who would think of using rocks such as green malachite or blue azurite to produce copper, or hematite for iron? At this point, we can only speculate. Perhaps malachite (which was used by the Egyptians as a cosmetic) was used to paint a piece of pottery, and transformed to copper in the kiln. The creation of metals from ores bearing no resemblance to the metals that could be extracted from them, marked the beginning of extractive metallurgy.

A bronze head produced by the lost-wax process.

Metal from Ores: The Bronze Age (3200 BC–1200 BC)

As most metal ores are compounds of the desired metal element with either oxygen or sulfur, some technique must be applied to free the metal from these other elements. The primary technique for millenia has been the use of carbon to draw out oxygen by forming carbon dioxide. Although this chemical theory was not known at the time, the processes by which metals could be extracted from their ores, were.

We take, as an example, the stunning transformation of malachite ($Cu_2CO_3(OH)_2$) to copper, which was performed by using charcoal both to provide the necessary heat, and to remove the oxygen. By setting layers of malachite between layers of burning charcoal and allowing the necessary heat to build, the carbon monoxide formed by the partially combusted charcoal will react with malachite, drawing out the oxygen as it forms carbon dioxide. As the process comes to completion, the malachite will have been transformed into copper. Such charcoal-fueled kilns could also reach the temperature required to melt copper (1,083°C), making it possible to pour the copper out into a mold, producing a *cast* copper form.

Whether it was originally developed from ores that also contained tin, or by means of willful experimentation of combining metals, mixing tin (or tin ore) with copper was discovered to produce a new substance, superior in every respect. This new material, *bronze*, was much stronger than copper, could be worked to a sharper edge, and melted at a lower temperature, making it easy to form cast* bronze objects.[6]

Many of the tools we use today—including the hammer, ax, chisel, and carpenter's rasp, were developed in the Bronze Age, as was the casting art known as the lost-wax process. In this technique, a wax model of the desired form to be cast in bronze is produced, with extra wax channels or guides (called *sprues*) added to it. This wax model is then coated in plaster or silica, which sets, and when it is baked, the wax melts out. This mold can then be filled with molten bronze, allowed to cool, and then the clay can be broken off, leaving the cast bronze object remaining. This technique is still used today for the casting of bronze sculptures.[7]

While copper could be found in the Mediterranean, tin could not, and the production of bronze required importing tin from trade routes stretching to what are today the British Isles, if not further.[8] The breakdown of these trade routes, and the lack of available tin, made the production of bronze impossible around 1200 BC.

---

4. As a metal piece is hammered, it gets stronger and stronger, and reaches a point at which further hammering will cause it to shatter, rather than bend. Heating the work-piece relieves internal stresses, and allows it to be further worked. This process is called "annealing."

5. Gold was mined in Egypt over 5000 years ago.

---

6. *Casting* means to pour liquid metal into a mold, into which shape it hardens. NB: terms marked with an asterisk (*) appear in the Glossary at the end of this section.

7. With the additional steps of the sculptor's clay work being coated in rubber, which is cut off, reassembled, and then filled with wax, which is then ready for the lost-wax process described here. See the video "Lost Wax Casting Process" by the National Sculpture Society: http://youtu.be/uPgEIM-NbhQ

8. Some evidence suggests that these trade routes extended to the New World.

Charcoal production: a century ago, and as recreated in modern times. The production of charcoal represented an important transformation. Wood, which contains many different chemical substances, is stacked in a large pile, covered with soil, and burned slowly in an oxygen-poor environment for a few days. The resulting product, charcoal, is almost entirely pure carbon. Charcoal is highly porous, allowing greater airflow in a furnace, and therefore higher temperatures and more rapid heating.

Above: Bronze casting. Molten bronze is poured into molds, where it hardens. Bronze melts at a lower temperature than copper and develops less air bubbles, making it easier to work with.

Left: A bronze sword. While copper was more durable than stone tools, it could not be made sharper than flint. Bronze is sharper than stone tools, and significantly stronger and tougher than copper.

Metal from Ores:
The Iron Age (1200 BC–)

The next great breakthrough in metallurgy was the introduction of a new metal source, known today to be the most plentiful metal in our planet's crust: iron. While iron requires greater temperatures and more extensive working than bronze in order to be as useful, this is more than made up for by its dramatically greater abundance.[9]

Iron was initially produced in a "bloomery furnace," in which iron ore and charcoal were heated together, producing carbon monoxide which removed the oxygen from the iron, as in the copper smelting discussed above. Heat for the chemical process was usually amplified by using a bellows to force more air into the furnace. This process did not reach temperatures great enough to melt iron, however, and the resulting bloom (known as "sponge" iron) had to be worked to remove the impurities, many of which did melt at these temperatures, and could be drawn out of the bloom by repeated hammering. After many cycles of heating and hammering, the bloom was sufficiently worked ("wrought") and relatively pure wrought iron* was the result. This labor-intensive process resulted in a product that could be formed into many shapes and whose ore was more plentiful than copper, yet could not be made as sharp as bronze, and was weaker.

Wrought iron implements were useful, but the production of steel was the advance that made iron a full replacement for bronze.[10] Steel* was made by the careful addition of carbon to wrought iron, by carburizing the surface of an iron implement by hammering it into charcoal, or by doing this repeatedly with iron sheets, until the whole material had become steel.

Modern Breakthroughs:
Into The Industrial Era

Blast furnaces, which forced hot air into the furnace column, were introduced in Europe in the twelfth century AD and reached temperatures hot enough to melt iron, producing *pig iron,** which contained a high level of carbon from the charcoal (or, later, coke) that it was surrounded by. This pig iron would then be worked in a finery forge (or later a puddling forge) to introduce oxygen to remove the carbon from the iron (inverting the initial smelting process).

---

9. While copper melts at 1,083°C, pure iron melts at 1,535°C, and cast iron objects (poured into molds from molten iron) were not produced in significant degree in Europe until the fifteenth century AD. In China, however, cast iron objects were made two millenia earlier, in the fifth century BC.

10. Today the use of steel is orders of magnitude greater than that of bronze.

Top: Eurico Zimbres, Bottom: Harvey Henkelmann

Hematite and taconite: two iron-containing mineral ores. Note how little they resemble iron in this form. Iron ores are much more plentiful on earth than are copper ores.

Morgan Riley

An iron "bloom" having the impurities beaten out of it by repeated hammering. This process is known as shingling. *Considering how much work is required, the name* wrought iron *is not such a mystery.*

A "damascus steel" knife blade, produced by repeated carburization and folding of wrought iron. The technique for producing damascus steel has been lost: exact replicas of this type of steel cannot currently be produced.

A major problem in the production of iron and steel was the intense use of charcoal: producing 10,000 tons of steel could require 100,000 acres of trees to be converted to charcoal in the Middle Ages.[11] Recall that wood, which burned at too low a temperature, and had too many impurities, could be converted to charcoal for steel production. Coal had the same problems as wood—too low a temperature and too many impurities. These problems with coal were solved by the brewing industry: it was purified in a way similar to that done with the transformation of wood to charcoal. Coal was burned in a low-oxygen environment to produce *coke*, in the same way that charcoal was produced from wood. This invention made possible the production of much more iron for society, and saved Europe's forests in the process. Even so, not all coal made coke that was acceptable for iron-work. Impurities in coal (particularly phosphorous) were not all removed in the coke-producing process, and only "metallurgical grade" coal was acceptable. In comparison, charcoal is almost completely pure carbon. The benefit of coke was not in its producing more heat, but in its being much easier to produce.[12]

The next major breakthrough was the use of the Bessemer process (invented in the middle of the nineteenth century), in which air was blown into melted carbon-rich

A blast furnace. These huge structures can tower over a hundred feet in the air, and continuously process enormous amounts of coke, iron ore, and flux to produce molten pig iron. In the production process, the iron picks up carbon from the blast furnace, which makes it quite strong, but it cannot be hammered or reshaped. It can be poured into molds as cast iron, but requires carbon removal to make steel.

A Bessemer converter, which was used to remove the carbon from high-carbon pig iron, by blowing air through the molten metal. The oxygen in the air combines with the carbon to form a gas, which escapes. This made for the beginnings of the modern steel era, by producing low-carbon iron much more cheaply than through the previous processes.

---

11. Iron production moved from county to county, or even nation to nation, based in significant part on the availability for forests to convert to charcoal. This figure comes from *Cathedral, Forge, and Waterwheel: Technology and Invention in the Middle Ages* by Frances and Joseph Gies, Harper-Collins, 1995, New York, N.Y. One cord of wood (transformed into charcoal) was required to process fifteen pounds of iron.

12. It did have one physical benefit: coke is stronger than charcoal, and does not compress or crumble as easily as charcoal when stacked in a furnace. This is important for allowing air to flow through the fuel and ore.

The Continuing Gifts of Prometheus

pig iron. The oxygen in the gas reacts with the carbon (and silicon) in the pig iron, producing more heat and allowing the process to continue without additional fuel. This process brought about a huge (nearly order-of-magnitude) reduction in the cost of steel, and its use expanded dramatically into applications that had called for wrought iron before. With the Bessemer process, steel was no longer produced by adding carbon to wrought iron, but could be produced from decarburized pig iron.[13] Again, the required carbon for steel (around 1%) required *adding* carbon to wrought iron (which had almost no carbon), or *removing* it from pig iron (which was 2–4% carbon).

Later advances in steelmaking built upon Bessemer's technique; easier-to-produce pig iron was to become the primary source of iron for steel, rather than wrought iron. The open-hearth furnace (used in the Siemens-Martin process) was similar to the Bessemer process, but operated more slowly, and used iron ore as an oxygen source, rather than air (which had the trouble of incorporating too much nitrogen into the metal). Because of its slower speed (some eight hours, rather than half an hour per Bessemer batch), open-hearth steelmaking could reduce the carbon content of pig iron to the proper amount for the desired steel, cutting the recarburizing step out of the process. That is, the Bessemer process required three steps: creating pig iron in a blast furnace, removing *all the carbon* by blowing air through molten pig iron, and then adding carbon again to produce steel; whereas the open-hearth process had only two steps: producing pig iron in a blast furnace, and removing the *appropriate amount of carbon* in the open-hearth to produce steel.

With a better control over the basic steelmaking process, and the longer process time of the open-hearth furnace, specific, "tuned" alloys of steel became an increasing portion of output. The most stunning example was the 1910s development of stainless steel, which contains a large amount of chromium (over ten percent) alloyed into the steel. This chromium forms a layer of chromium oxide film, which protects against the corrosion and rust that would otherwise eventually destroy steel. Because the chromium is mixed into the steel, rather than just coating it,[14] scratches and dents are self-healing: the newly exposed steel also contains chromium which quickly oxidizes, forming a new protective layer. Other common alloying metals are nickel, molybdenum, and manganese. Like per capita energy use (by source), which serves as an important physical economic indicator of development, stainless steel consumption per capita reflects high-technology economic activity.

New York City's Chrysler Buildling. Completed in 1930, the top of the building was wrapped in stainless steel. Scratching and abrasion do not damage the corrosion resistance of this steel.

The "spangled" appearance of a galvanized steel handrail. The thin zinc coating, less than a millimeter thick, protects the steel from corrosion, but will wear off over time, and can be scratched or abraded off.

Today, the open-hearth furnace has been replaced by basic oxygen steelmaking (BOS, developed in the middle of the twentieth century) which is quite similar to the Bessemer process, but uses pure oxygen (unavailable in the needed quantities in Bessemer's day) rather than air. Using turboexpander-generated liquid oxygen, BOS steelmaking can transform pig iron into steel in a fraction of the time required by the open-hearth furnace, and is the primary method of producing steel today, capable of operating on metal scrap and pig iron. Advances in chemistry and spectroscopic instrumentation make it possible for BOS steelmaking to stop at just the right point, when the desired carbon level is reached, even though the process occurs quickly.

Nearly one third of steel production comes from recycling scrap metal in electric arc furnaces (EAFs), which pass an electric current through the metal, directly heating it in the process.[15] This electricity-intense process consumes around 400kWh of electricity per ton of steel, and can be easily scaled down for small batches of specialty steel. Obviously, a fully nuclear economy and the cheaper electricity (and process heat for pre-heating) would make electric arc furnaces much cheaper (physically) relative to coke-fired blast furnaces.

---

13. On top of the coke already required to produce iron, the process of adding carbon to wrought iron to produce steel required several additional tons of coke per ton of steel produced.

14. *Galvanized* steel, which is coated with a thin layer of zinc (producing a characteristic "spangled" appearance) is also protected against corrosion, but scratches that penetrate the thin zinc coating will cause the rusting away of the steel.

15. EAFs account for 29% of steel production. This stunning process is worth seeing! One example video: http://youtu.be/G6Uxh-xtU-g

Fully replacing blast furnaces and the use of coke requires another breakthrough. Even though EAFs can reach great heats, the other aspect of ore processing is *reducing* it chemically: removing the oxygen to which the iron is bound. This is the *chemical* role of coke in the blast furnace, in addition to its heating role. An EAF cannot perform this chemical reduction, and is therefore only useful, at present, for processing metals, but not ores.

In an economic platform capable of large-scale deployment of *plasma torches*, the reduction of ores could be performed without using coke at all, as the chemical change can be brought about directly, without carbon to bond with the oxygen.[16] The metal would still need to be separated from the oxygen, which currently occurs by a phase change (producing carbon dioxide gas), but which could be performed by ionizing the metal and separating it with a magnet, before it cools out of its plasma state and recombines with the oxygen. With such technologies, multiple processes could be combined into one: the coke ovens required to produce coke, the blast furnaces to produce pig iron, the refining processes to remove carbon, and even the totally different technology used to produce aluminum, could all have their tasks performed by such a "universal machine," operating with plasmas.

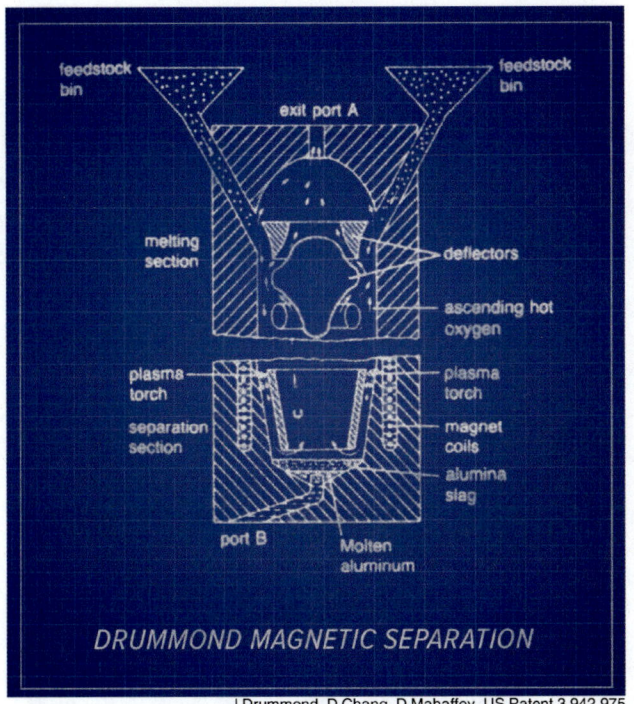

A design for a magnetic separation machine for ore processing. This diagram indicates its role for aluminum production, but the same basic design would function for iron processing as well.

16. See article on plasma torches in "Nuclear NAWAPA XXI: Gateway to the Fusion Economy," available at:
http://21stcenturysciencetech.com/Nuclear_NAWAPA.html

Metallurgy in the Modern Era: the Future of Metals and Metallurgy

Until the recent two centuries almost the sole use of metals discussed here had been for structural, rather than specifically chemical use. The characteristics of the metal that were sought out were *physical* properties, such as strength, flexibility, hardness, density, and ductility. Advances in chemical understanding gave new uses to metals and alloys, and with the advent of the electrical era, entirely new characteristics of metals became important. The employment of electric motors, rather than steam engines in factories, required metals that were economical, workable, and conducted electricity well. The excellent electrical conductivity of copper (exceeded only by silver) and its flexibility makes it the primary metal used for building wiring. Where weight and cost are an issue (as in high-voltage transmission lines), aluminum is used.

The recent few decades' change in the applications of electricity has brought to life previously unconsidered properties of metals and similar elements. The invention of the transistor in 1947 marked the beginning of the intense use of semiconductor materials, in which the electrical properties of silicon (a metalloid) are engineered by the incorporation of other elements, in order to bring out very specific electrical properties. Computer-automated control of machining and industrial processes was possible on a large scale with the development of semiconductor integrated circuits. Relatively rare metals, whose structural characteristics in alloys are sometimes impressive, are increasingly being used for their chemical and electrical characteristics, serving specialized roles as chemical catalysts, phosphors, magnets, and motors.

And now, the engineering and science breakthroughs made possible by these increasingly precise components open the way to "universal machines" operating with plasmas and magnetic separation, machines which can fundamentally transform the way metals, both common and rare, are formed into useful products. These changing roles of metals are a specific case of the discovery of new types of physical principles, as brought about by the creation of modern chemistry and electrodynamics.

In the next section of this report, we move from physical properties of materials to the chemical nature of matter, which brought a new understanding of metallurgy, and opened the way to the next dimension of physical actions by man. Modern chemistry brought together many different chemical properties, to develop the unity of matter expressed by Mendeleev in his periodic table of the elements.

Metallurgy Glossary

Ore Reduction

In the chemical sense of *reduction*, this refers to separating the metal element (e.g., copper or iron) from the oxygen to which it is bound. Coke or charcoal act as reducing agents, binding oxygen in the ore to their carbon and producing carbon dioxide in the process, which escapes as a gas.

Casting

When molten metal is poured into a mold and allowed to cool into a desired shape, this process is called *casting*, and the resulting object is *cast*, such as cast bronze, or a cast iron frying pan. Cast iron objects (produced from the material known as pig iron) have high levels of carbon, which makes them strong but very brittle. Cast iron cannot be worked by a blacksmith, even when heated. Hammering it would break it, rather than bend it.

Wrought Iron

Iron produced in a bloomery furnace (or another process that does not cause it to actually melt) contains many impurities. As the resulting hot iron bloom is hammered, slag is worked out, and when the iron has been sufficiently worked in this manner, it is said to be wrought (an old form of "worked"). Wrought iron can be hammered and worked by a blacksmith into desired shapes, although it is not very strong. Wrought iron is no longer produced on a commercial scale today, and many of its former applications are now met with steel. Wrought iron has a low carbon content.

Pig Iron

Iron ore that has been melted in a furnace, picking up excess carbon along the way, containing 2–4% carbon. The resulting "pig iron" gets its name from the shape molten iron would take when poured out into sand molds: the central runner and side ingots resembled a mother sow feeding her piglets. Molten pig iron can be cast into molds to form cast iron, or further processed to remove carbon for steel production.

Steel

Steel is iron that has a specific carbon content (around 1%), giving it both strength and workability. Wrought iron is workable but weak, and cast iron is strong but cannot be reshaped. Steel combines beneficial characteristics of both materials, and has almost completely replaced wrought iron, although cast iron still finds applications.

Working in a steel plant. Protective clothing keeps employees safe as they work with molten metal at temperatures exceeding 1500°C.

Chemistry:
The Active Power of the Elements

by Michael Kirsch

Dmitri Mendeleev, in his great work, *The Principles of Chemistry*, fundamentally transformed the way man understands the material world around him, by introducing a new concept of fundamentally *chemical*, rather than *physical* principles. Just as the economic value of any product or process depends on the context in which it exists, so too do the physical properties of all materials change: from density to specific heat, from electrical properties to color. Mendeleev went beyond the measurable *physical* properties of the compounds that elements enter into, to discover the periodic ordering of the *chemical* properties of the elements: the chemical transformations they were capable of. Allowing nature to speak for itself, he discovered a unity underlying all of matter, and swept away the foolishness of the alchemists and reductionists.

Setting the Stage

In the middle of the 15th century, Nicholas of Cusa asked what benefit man would gain if weight scales were used to compile the weights of metals, plants, and many other things, in order to measure the unseen in things which cannot be sensed directly.[1]

A fundamental turning point in the process of revealing the lawfulness of the chemical structure of the universe was achieved when, in the 18th century, Antoine Lavoisier subjected the transformations of substances to weight scales. Among the experiments he performed, Lavoisier found that by weighing green powdery copper-carbonate (malachite) before and after heating it in a container of air, the weight of the resulting black substance was less than the original green substance, meaning that something had separated from it. He noticed that a gas was released during the heating, and by funneling it to exit through a tube of the heated vessel, he could measure the quantity of the released gas, carbon dioxide (CO_2). Since the sum of the weights of the new black copper oxide (CuO) and newly formed carbon dioxide gas was equal to the former weight of the copper carbonate originally taken, Lavoisier was led, by this and other demonstrations, to the law of the indestructibility of matter: that in all transformations of compounds of elements into others, matter is not created and does not disappear, but that "the sum of the weights of the substances formed is always equal to the sum of the weights of the substances taken."[2]

Coinciding with this discovery, Lavoisier was able to conduct other investigations which revolutionized the conception of substances altogether. After heating me-

Dmitri Mendeleev (1834–1907), whose discovery of the periodic ordering of the elements, which provided a universal view of all matter, is left out of today's education, or is shamefully glossed over. Working through his discovery should be required for any student of chemistry.

---

1. Cusa, for example, in his *Idiota de Staticis Experimentis*, proposed to come to the nature of herbs by using weight and taste, rather than taste alone, and to measure the sickness of a man by his blood quality, using its color and weight, rather than using the sense of sight alone; on the basis of the agreement or difference of weights of these substances, the correct dosage of herbs could be given for the correct illness. Among the tasks to be taken up were weighing the amount of water displaced by different submerged metals as a means to measure and determine their non-visual differences, measuring the invisible power of a magnet by how much weight it displaced on a scale, and weighing of seeds taken from different regions to measure the power of the sun at different latitudes.

2. Mendeleev was explicit that all progress in chemistry had been based on Lavoisier's discovery of this fact, since by applying the indestructibility of matter, it was obvious whether one of the resulting substances was being overlooked, as the consequent weight would come out unequal.

tallic mercury in a sealed vessel of air for twelve days, he noted that red scales formed on the mercury. After weighing the remaining air in the vessel, he found that it had decreased in weight by the amount the mercury had increased in weight. He discovered that the mercury had taken in a life-supporting gas, oxygen, forming mercury oxide, and leaving behind another gas in the vessel. This other gas (nitrogen) did not support life, leading to the revolutionary discovery that air is not its own element. Such experiments led him to discover that many compounds could be reduced to simpler states, but only up to a point, writing that if "we associate with the name of elements, or of the principles of substances, the idea of the furthest stage to which analysis can reach, all substances that we have so far found no means to decompose are *elements* for us."

Mendeleev's Concept of the Element

A century later, Mendeleev drew out the implications of the indestructibility of matter further, making more explicit the fact that the subject under investigation was not one of sense perceptible substances, but certain characteristics which *cause* change but are themselves unchanged: "many elements exist under various visible forms while the *intrinsic element* contained in these various forms is something which is not subject to change." Mendeleev spelled out this difference between sensible forms and the true conception of elements in detail.

Making use of charcoal as a case in point, he stated that although the matter making up charcoal is found in organic substances in combination with hydrogen (H), oxygen (O), nitrogen (N), and sulfur (S), "in all these combinations there is no real charcoal, as in the same sense there is no ice in steam. What is found in such combinations is termed 'carbon'—that is, an element common to charcoal, to those substances which can be formed by it, and also to those substances from which it can be obtained." Similarly, carbon appears uncombined with other elements in charcoal, graphite, and diamond, but yet "the element carbon alone contained in each *is one and the same.*"[3] Carbon dioxide contains carbon, and not charcoal, or graphite, or diamond. Therefore, when iron ore or another metal oxide is burned with charcoal and the heat allows the oxygen from the metallic oxide to combine with the carbon in the charcoal, it is not *charcoal* which is forming a compound with the oxygen of the metal; the material expression of carbon in the newly formed carbon dioxide is incommensurable with that in charcoal.

For Mendeleev, an element was known not by its physical characteristics, but by the chemical transformations it can undergo. He wrote: "Mercury oxide does not contain two simple bodies, a gas and a metal, but two elements, mercury and oxygen, which, when free, are a gas and a metal." The essences, or tranformative potential of elements, are in compounds, rather than any particular state of the element. In this way, the presence of the principle is made primary, or as he wrote, "the composition of a compound is the expression of those transformations of which it is capable." Similarly, there is not oxygen of any one form in oxygen gas, ozone, water, nitric acid, or carbon dioxide, but a principle which is capable of producing all of them, leading to the truth that, "as an element, oxygen possesses a known chemical individuality, and an influence on the properties of those combinations into which it enters."

Mendeleev did not make the transformations of sense perceptible materials the study, but rather *the power to transform*—the invisible principles which characterize and determine possible actions, which are maintained through all the visible changes of compounds. This is further underscored in an 1889 speech, where he referenced this conception:

> Before there was an idea of a primary matter, as to the material world, they adopted the idea of unity in the formative material, because they couldn't resolve any other possible unity to connect the relations of matter. I have discovered through the universe a unity of plan, a unity of forces, and a unity of matter; and the convincing conclusion of modern science compels everyone to admit these kinds of unity. We must explain the individuality we see everywhere. It has been said of old [by Archimedes], "Give a fulcrum, and it will become easy to displace the earth." So also we must say, "Give anything that is individualized, and the apparent diversity will be easily understood." Otherwise, how could unity result in a multitude?[4]

The Principles of Chemistry

As with Kepler's *Mysterium Cosmographicum*, which Kepler, upon its re-publication, decided not to change but only update with footnotes, Mendeleev never changed the presentation of his original 1869 *Principles of Chemistry*, but continuously updated the old version with added footnotes, which by the seventh edition were as long as the original book itself. Mendeleev's faithfulness to his original presentation reveals that he considered the key steps of his breakthrough, as originally presented, to be a correct direction of scientific thought.

For the first nine chapters of *Principles of Chemistry*, Mendeleev investigates the properties of the four elements hydrogen (H), oxygen (O), nitrogen (N), and car-

---

3. Mendeleev, *Principles of Chemistry*, George Kamensky, trans., New York: Longmans, Green, and Co., 1891.

4. Mendeleev's memorial Faraday Lecture to the Chemical Society in London.

bon (C), and the simple patterns of the way in which other elements combine with them, these compounds serving as *types* for compounds of other elements. For example, the number of atoms of hydrogen which entered into molecules per one atom of another element—be it one, two (as with oxygen in water, H_2O), three (like ammonia, NH_3), or four (with carbon to form methane, CH_4)—made it possible to foretell other compounds these elements could form.

But, it is not possible to foretell all properties from this alone, and in Chapter 10, he turns to a deeper characteristic which leads him to then begin discussing the breakthrough regarding a system of organization of the elements as a whole. He says that there exist among the elements qualitative analogies and relations which are not exhausted by their compounds, but are most distinctly expressed *in the formation of bases, acids, and salts of different types and properties*, and that for a complete study of the nature of the elements, it is especially important to become acquainted with the salts. Certain elements provided extreme examples of the actions that others are capable of performing.

At one end, chlorine provided a unique example. "It forms strong acids with hydrogen and oxygen-acids that give salts, such as common table salt, upon combining with metals."[5] Four elements, fluorine, chlorine, bromine, and iodine, had these same properties of reaction when combining with metals and non-metals. This group was called the halogens ("salt producers," using the Greek word for salt), elements which all gave their compounds specific properties which they alone shared.

Foremost among their properties is the mentioned salt-forming *acid oxides*. The acid oxides of bromine and iodine, are similar to the acid oxide of chlorine, as hypobromous acid (HOBr) corresponds in its properties with hypochlorous acid (HOCl), both formed by adding pure bromine or chlorine to water. The salts of these acids,[6] such as sodium hypochlorite (NaOCl) both share the same bleaching property and are also both very unstable.

At the other end, Mendeleev then introduced sodium (Na) and its analogues, known as the *alkali metals*, which are characterized by their power to form the most *basic* oxides but no acid oxides. As he wrote later, "…the sodium contained in table salt, NaCl, is the model for elements giving only bases, and not oxygenated acids. In its combination with oxygen, it gives a base, sodium oxide." Sodium has a power of decomposing water easily through its capacity to form the most stable basic oxides. It has such an affinity for oxygen that it is not found naturally, but oxidizes almost immediately when exposed to air. Other unique characteristics of sodium are its power to form salts that are soluble, like sodium sulphate and sodium carbonate.

Other elements also do not appear in a free state, oxidize in air quickly, decompose water, form soluble hydroxides, and form similar salts. These are potassium, lithium, and cesium, known collectively as the "alkali metals."

By means of comparison with the halogens and alkalis, other elements can be considered with regard to these extremities. Some elements approach the alkali metals in capacity of forming salts and not acid compounds, but are not as energetic as akalis. Other elements approach the halogens, but do not have the same energy: in a free state they combine with metals easily, but do not form salts like halogens. Sulfur, phosphorous and arsenic fall here. Then there are elements which are neither like alkali metals nor halogens, such as carbon, nitrogen, and oxygen.

Prior to Mendeleev, some had ordered the elements according to their atomic weights,[7] but suffered from faulty pure numerical orderings, and had groups with completely dissimilar elements listed together. Mendeleev did something different:

> Nobody has established any theory of mutual comparison between the atomic weights of unlike elements although it is precisely in connection with these unlike elements that a regular dependence should be pointed out between the properties and the modifications of atomic weights.

He elaborates:

> Everybody understands that in all changes in the properties of simple substances, *something* remains unchanged and that, in the transformation of the elements into compounds, this material something determines the characteristics common to the compounds formed by a given element. In this regard only a numerical val-

---

5. "How I discovered the Periodic Law" 1899, in *Mendeleev on the Periodic Law: Selected Writings, 1869–1905*, Dover Books on Chemistry, William Jensen, editor, 2005.

6. An acid, saturated with an alkali solution, will release heat, and if evaporated, a solid crystalline substance is yielded. This is called a salt, in the chemical sense, a compound of definite quantities of an acid with an alkali. More generally, a salt is an acid in which hydrogen is replaced by a metal. In this case the hypochlorous acid HOCl, becomes sodium hypochlorite, NaOCl, a salt, where hydrogen (H) is replaced by sodium (Na).

7. Atomic weight can be understood as follows: Forming water chemically, from hydrogen and oxygen gas, reveals that there are about 8 parts oxygen to 1 part hydrogen, by mass. However, Joseph Gay-Lussac had earlier found that two volumes of hydrogen gas combine with one of oxygen to form water; therefore, there are two volumes of hydrogen making up the 1 part to 8 of mass. In other words, there will be two hydrogen atoms for each oxygen atom which make up the water molecule. The atomic weight of the two elements was therefore 1 to 16, oxygen having atomic weight of approximately 16, in relation to hydrogen taken as 1. This measurement is a chemical property, not a physical one: it can be discovered only by chemical processes of combination.

ОПЫТЪ СИСТЕМЫ ЭЛЕМЕНТОВЪ

ОСНОВАННОЙ НА ИХЪ АТОМНОМЪ ВѢСѢ И ХИМИЧЕСКОМЪ СХОДСТВѢ

```
                    Ti = 50    Zr = 90    ? = 180.
                    V = 51     Nb = 94    Ta = 182
                    Cr = 52    Mo = 96    W = 186.
                    Mn = 55    Rh = 104,4 Pt = 197,4.
                    Fe = 56    Ru = 104,4 Ir = 198
               Ni = Co = 59    Pl = 106,6 Os = 199.
H = 1               Cu = 63,4  Ag = 108   Hg = 200
       Be = 9,4 Mg = 24  Zn = 65,2  Cd = 112
       B = 11   Al = 27,4 ? = 68    Ur = 116   Au = 197?
       C = 12   Si = 28   ? = 70    Sn = 118
       N = 14   P = 31    As = 75   Sb = 122   Bi = 210?
       O = 16   S = 32    Se = 79,4 Te = 128?
       F = 19   Cl = 35   Br = 80   I = 127
Li = 7 Na = 23  K = 39    Rb = 85,4 Cs = 133   Tl = 204
                Ca = 40   Sr = 87,6 Ba = 137   Pb = 207
                ? = 45    Ce = 92
              ?Er = 56    La = 94
              ?Yt = 60    Di = 95
              ?In = 75,6  Th = 118?
```

Д. Менделѣевъ

Mendeleev's 1869 periodic table. The relationship between groups of elements that are similar in their chemical activity, and the atomic weights of those elements, gives rise to the periodic relationship expressed in the form of a table by Mendeleev. Each column is a period. Note the question marks for elements that had not yet been observed, but that Mendeleev had hypothesized to exist.

ue is known, and this is the atomic weight appropriate to the element. The magnitude of the atomic weight, according to the actual, essential nature of the concept, is a quantity which does not refer to the momentary state of a simple substance but rather belongs to a material portion of it—a portion which it has in common with the free simple substance and with all its compounds. The atomic weight does not belong to coal or to diamond but rather to carbon.[8]

---

8. "On the Correlation Between the Properties of the Elements and their Atomic Weights," Mendeleev 1869, in *Mendeleev on the Periodic Law: Selected Writings, 1869–1905,* Dover Books on Chemistry, William Jensen, editor, 2005.

Periodic Properties

Mendeleev writes: "The formation of such natural groups as the haloids, the metals of the alkalis...and alkaline earths...furnished *the first opportunity* of comparing the different properties of the elements with their atomic weights."[9]

By comparing the propensity for combination with atomic wieghts, certain powers of transformations became most important. Mendeleev saw that the halogens arrange themselves by their physical properties, such as ease of oxidation and the stability of the oxides they formed, in the same order as they stand in respect to their atomic weights. Their atomic weights are fluorine (F)=19, chlorine (Cl)=35.5, bromine (Br)=80, and iodine (I)=127. Accordingly, iodine acid oxide is more stable than chlorine acid oxide, with iodine having a much greater affinity for oxygen than chlorine. Mendeleev excitedly points out, that bromine, whose atomic weight is nearly halfway between that of chlorine and iodine, also holds an intermediate position with respect to oxide stability. Fluorine, he says, because of chlorine's difficulty in doing so, predictably does not form an oxide at all.

Their relation to hydrogen can also be so compared, only in reverse order. Fluorine has such an affinity with hydrogen that it decomposes water at room temperature, while iodine has an enormous difficulty in combining with hydrogen. Their compounds with hydrogen are therefore likewise arranged according to atomic weight, with hydrogen chloride being the most stable, hydrogen bromide occupying the middle position, and hydrogen iodide the least stable. Other properties corresponded with atomic weight as well: the higher the atomic weight, the higher the specific gravity, vapor density, and melting and boiling points.

The case is similar for the alkali metals, whose atomic weights are lithium (Li)=7, sodium (Na)=23, potassium (K)=39, rubidium (Rb)=85, and cesium (Cs)=133. The chloride salts of lithium and sodium are soluble, but the chloride salts of potassium, rubidium, and cesium are hardly soluble. Thus, the greater the atomic weight, the less soluble is the salt. The variation of properties with the weight even shows itself in the free metallic form of the metals themselves, not just their salts; lithium volatilizes with difficulty, while sodium volatilizes by simple distil-

---

9. "On the Periodic Regularity of the Chemical Elements," Mendeleev 1871, *ibid.*

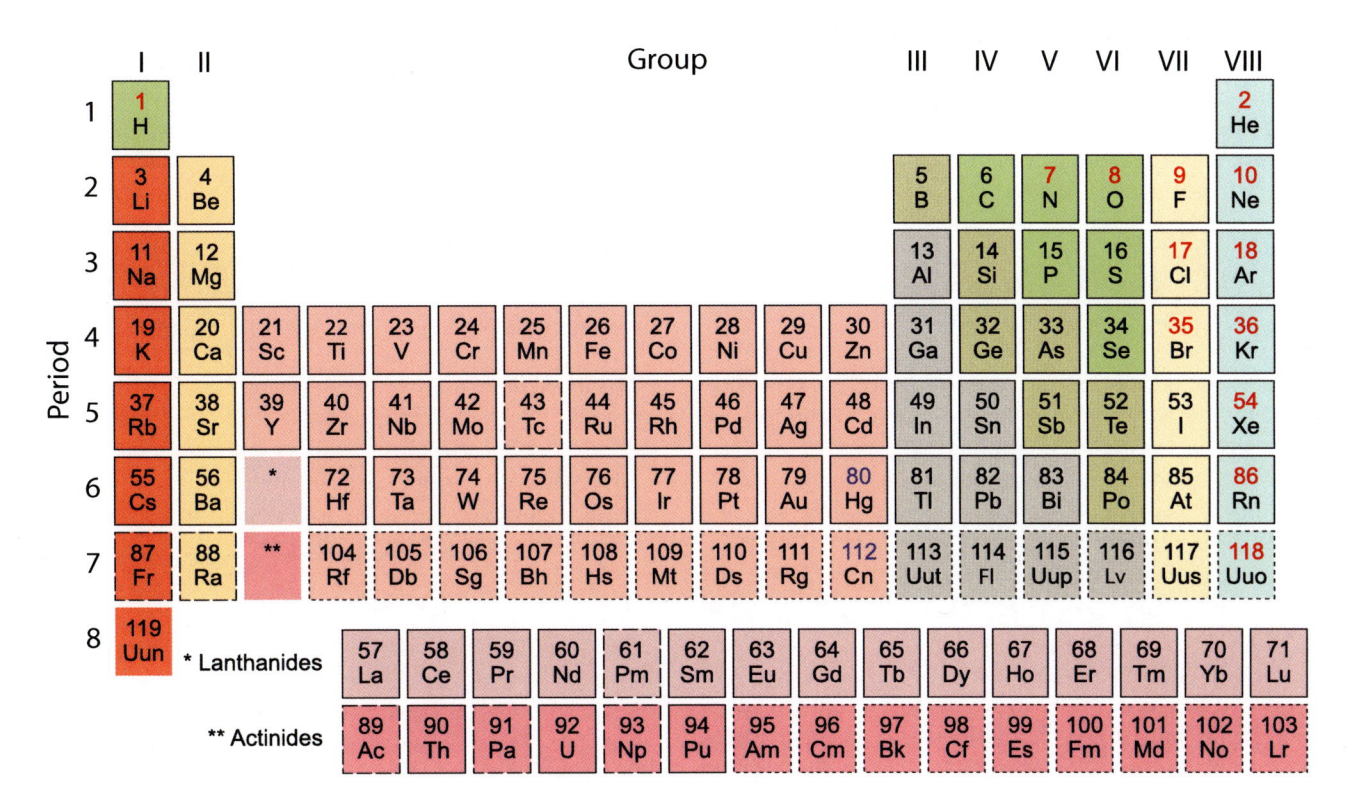

A modern table of the elements, developed as an outgrowth of Mendeleev's periodic ordering. The presentation of this table to chemistry as a given, without developing the process of experimentation and thought that gave rise to Mendeleev's concept, has stunted their creative potential.

lation. Potassium volatilizes yet more easily than sodium, and rubidium and cesium are still more volatile. In other words, ease of volatility increases with atomic weight.[10]

This method thus became a measurement to determine whether a grouping was real (intrinsic to the elements themselves), or merely a false imposition. In an account of an earlier attempt at organization, Mendeleev found that of the characteristic features of alkalis and halogens, *their extreme basic and acid oxides*, formed the extremes of a periodicity of types of basic and acidic properties.

Li 7   Be 9.4   B 11   C 12   N 14   O 16   F 19
Na 23 Mg 24   Al 27.3 Si 28   P 31   S 32   Cl 35.5

For example, moving from the right to the left, in relation to hydrogen, the acidic character lessens. Thus hydrochloric acid (HCl) is a very decided acid of great stability; whereas hydrogen sulfide (H_2S) is a weak acid decomposed by heat, and phosphine (H_3P) has almost no acidic properties. In relation to oxygen, the case is inverted: moving from left to right, the oxides, starting with sodium oxide (which is so stable that it only separates with oxygen upon being heated to the temperature of melting iron) decrease in stability.

In addition, the ability of elements to combine in ways similar to hydrogen, oxygen, nitrogen, or carbon, now correlated exactly with this ordering of the atomic weights. The numerical relationships of combination were known as valence numbers. The fluorine and lithium groups (halogens and alkalis) had valence 1, combining like hydrogen. The beryllium and oxygen groups had valence 2, the boron and nitrogen groups valence 3, and the carbon group valence 4. Corresponding members in different rows (such as lithium and sodium) produce the same types of compounds: they possess the same valences.

Such characteristics as these created boundaries which were used in placing the rest of the elements within the ordering system and redefined earlier known laws of chemistry from a higher standpoint. Seeing the correspondence of the atomic weights with such properties as these guided the organization of the system. After these chapters, in which his principle was applied, came the formalization of his periodic law:

> The properties of the elements (and of the simple and compound substances which they form) show a periodic dependence on their atomic weights.... All of the

---
10. Also, in another group, the "alkaline earth" group—Be, Mg, Ca, Sr, and Ba—the alkaline properties increase with atomic weights, and show themselves in many of their compounds. Ca decomposes water with ease, Mg does with difficulty, and Be not at all.

The Continuing Gifts of Prometheus

functions which express the dependency of the properties on the atomic weight may be characterized as periodic functions. At first, the properties of elements change as the atomic weight increases; then they repeat themselves in a new series of elements—a period—with the same regularity as in the preceding series.

Therefore, what defines an element? Is it its material form? No, it is defined by how it is situated in a periodic set of relationships to all others.[11] From that standpoint, it is important to supersede common blunders, which reduce his breakthrough simply to an organization of the elements according to their atomic weights. Others had done that. Mendeleev allowed for these individualities—investigated as characteristics of change, present in the smallest part, which influence any substance dynamically—to define themselves and organize themselves by their unique actions, all in relation to another chief characteristic, atomic weight.

We highlight here the notable fact that just as Cusa had proposed to compare the measurement of the weight of blood with the visible color of the blood, to get at a truth by relating the two, for Mendeleev, it was the relation between, on the one side, these certain characteristic actions, such as the power of their acid and basic oxides, how they combined with other elements—characteristics that formed groups of elements—and, on the other, the invariant of *atomic weight*, which revealed the unique periodicity and presence of a higher principle of organization, one not completed by Mendeleev.[12]

Mendeleyev gave this explanation of the standpoint from which he discovered the periodic law.

> Of the exact nature of matter we have no knowledge.... We are unable to comprehend matter, force, and the soul in their substance or reality, but are only able to study them in their manifestations in which they are invariably united together, and beyond their inherent indestructibility they also have their tangible, common, peculiar signs or properties which should be studied in every possible aspect. The results of my labors in the study of matter show me two such signs or properties of matter: (1) the mass which occupies space and evinces itself in gravity or more clearly and really in weight, and (2) the individuality expressed in chemical transformations and most clearly formulated in the notion of the chemical elements. In thinking of matter outside any idea of material atoms, it is impossible for me to exclude two questions: How much and what kind of matter? Which qualities correspond to the conceptions of mass and of the chemical elements? There the thought involuntarily arises that there must be some bond of union between mass and the chemical elements; and as the mass of a substance is ultimately expressed (although not absolutely, but only relatively) in the atom, a functional dependence should exist and be discoverable between the individual properties of the elements and their atomic weights.[13]

In Conclusion

In approaching a new, undiscovered principle, the wise thinker, since the days of Nicholas of Cusa, always chooses to let the higher process define itself.[14] Rather than describing a new process by its effects, the human mind must always get "inside" the higher physical process that is being investigated and let it define its own laws; not by what it produces, but by investigating *how* it produces it.

Through the contributions of Cusa, Lavoisier, and Mendeleev to chemistry, we see new dimensions of characteristics of matter and its actions. We find specifically chemical properties of the elements themselves, distinct from the physical properties of the compounds they enter into. This chemical understanding shed new light on processes of the past—knowing why they occurred as they did—and made it possible to hypothesize new technologies and experiments in the future. Mendeleev opened a new understanding, a new dimension of matter itself, one that forms the basis of much of what has come since: from petroleum refining to photography, from pharmaceuticals to batteries, to the hundreds and thousands of other new chemical compounds developed since his time, and those still to be invented.

The next dimension of physical chemistry to explore is the domain of electromagnetism.

---

11. It may also be noted that while the periodic law showed that "our chemical individuals display a harmonic periodicity of properties, dependent on their masses," Mendeleev made the point later that most periodic functions are continuous, but the one which he discovered is peculiarly made up of discrete jumps, in addition to various lengths of the periods. It is notable that therefore, mass is capable of a non-linear function in that it does not have a continuous relationship to the chemistry of matter. It is periodic. As the mass of elements increases discretely, so properties change, and then at another discrete mass change, it cycles back in terms of the properties, but slightly changed.

12. For a more in-depth demonstration of the periodic law, readers are referred to Chapters 15 on that subject in Mendeleev's *Principles of Chemistry*.

13. Chapter 15 of Mendeleev, *Principles of Chemistry*, George Kamensky, trans., New York: Longmans, Green, and Co., 1891.

14. This approach rests on the foundations of the method of modern science, in line with Nicholas of Cusa's approach to the quadrature of the circle, Kepler's and Leibniz's method of dynamics and higher transcendentals, up through Gauss and Riemann's elliptical functions. See Kirsch, "The Calling of Elliptical Functions," *Dynamis* magazine, December 2008, at: http://science.larouchepac.com/publications/dynamis/issues/december08.pdf

Electromagnetism
A New Dimension

by Creighton Cody Jones

The discovery of the principles of electromagnetism is a provocative slice of the history of human evolution. For man, the process of evolution is not expressed as some biological change, but rather as a willful change in his relationship to the universe, as a function of the discovery of new universal principles. In other words, man intentionally evolves himself through the discovery, transmission, and assimilation of principles of science, to the effect of gaining greater control over the world around him, creating new states of order that had not previously existed. This is most certainly the case with the successive self-feeding process of discovery and application of the many aspects of electromagnetism.

The discovery of electricity was not, as some initially thought, the discovery of some new motive fluid, but was rather a discovery of a hidden fundamental characteristic of all the elements of the periodic table. Today it is understood that electricity is the propagation of a specific form of action that requires a material carrier, and is intimately associated with a fundamental and inherent property of all matter, called *charge*.

This understanding, that chemical elements are composed of fundamental charge carriers, was not fully realized until after 150 years of modern experimentation with, and economic application of, electricity. This understanding came with the discovery of the electron by J.J. Thompson in 1897, where already at that time the electrical field concept was under full development, the fruit of the labors of such scientific luminaries as Ampère, Gauss, and Weber. And it is from these foundations that mankind has progressed to having the ability today of near-instantaneous communication across oceans, and to set foot on the moon.

It was, and continues to be, a discovery of a new type of Promethean fire, sparked by an investigation into the deeper fabric of physical chemistry. Whereas the fire of metallurgy was seated in burning embers, the fire of electromagnetism starts with the rubbing of amber.

The Evolutionary History of Man

On Earth, the different acts of producing, harnessing, moving, and using electricity call upon different electrical characteristics of materials. Some produce electricity by being rubbed, vibrated, or heated, like the static charge built up on rubbed amber. Some are excellent transporters of electricity, like copper, while others prevent its motion, as seen with glass.

These general properties, however, can be changed when materials are exposed to extreme or rare conditions.[1] For example, among ceramics, which normally act as insulators that restrict the flow of electricity, some express the opposite property at extremely low temperatures, becoming superconductive. Carbon, which, in a diamond crystal lattice is completely non-conductive, also becomes superconductive when its structural configuration is changed, in the form of graphene—a one atom thick lattice of carbon.

In essence, how we choose to use electricity, or how we can use electricity, is a function of the varied responses of different elements under different conditions to the interactions with moving charge and charged substances. Thus, the study of electromagnetism is very much an investigation into the deeper domains of physical chemistry.

All elements are composed of charge, but we often find them in states where charges are balanced between oppositely charged electrons and protons, and therefore they do not express, on the macroscopic scale, any immediately measurable electric effect, as with a typical piece of wood or plastic. However, there are numerous conditions where many elements and their composites do express electrical effects, as we will see.

Before we get to the more modern development of the battery and the discovery and application of flowing current, it is important to briefly discuss the form of electricity as it was explored by Ben Franklin and others in the early to mid-18th century.

At this time, man was investigating a phenomenon that had been known as far back as the ancient Greeks. It was found, that when a substance such as amber was rubbed with fur, it could be made to attract pieces of lint or hair.[2] Other substances were found that exhibited this prop-

---

1. More often than not these "rare" conditions are man-made conditions.

2. The term *electricity* comes from the Greek word for amber (*elektron*).

erty when rubbed, and it was discovered that there were two different and opposite kinds of electricity: that produced from rubbing amber (*resinous* electricity) and that produced by rubbing substances such as glass (*vitreous* electricity). Either amber or glass would attract small bits of metal, but once a piece of metal had touched charged glass, it would be repelled by it, but would be *attracted* to the amber.

It was the buildup of static charge and its discharge that was at the core of many of Franklin's experiments, and is most famously displayed by his use of Leyden jars. These devices were essentially closed glass jars with a metal rod poking out of one end of the lid while in the other direction the rod extended into the open space of the jar. A charge would build up inside the jar when the exposed end of the rod was rubbed with a charged piece of amber, glass, or other material, as if to wipe the charge onto the metal rod. Franklin considered the two types of electric charge not to be two different fluids, but rather an overabundance or a lack of a common electrical fluid. His terminology for "positive" and "negative" electricity is still used today.[3]

It is now understood that in one case there is a buildup of excess electrons on the surface of the substance, producing what Franklin had arbitrarily dubbed "negative" charge, or in the other case a "positive" charge is built up through the removal of electrons. What is created is a static charge, whose intensity is a function of the amount of charge built up and contained by a given region. It is akin to the buildup of pressure which results from pumping more and more gas into a rigid container. The full charge which has been built up is then discharged when the metal rod is brought into contact with a conductive body, such as a finger or a wire.

Various technologies were developed to exploit the new understanding of this phenomenon, including attempts at a telegraph, but the difficulty in making a signaling device with static electricity, and the amount of effort required to produce the electricity, made the technology impractical. The Leyden jar and the various electrostatic devices did, however, provide for the means for experimenting with electricity in a variety of contexts, including in biological experiments, which then gave way to the development of the battery and the sustained flowing current of electricity.

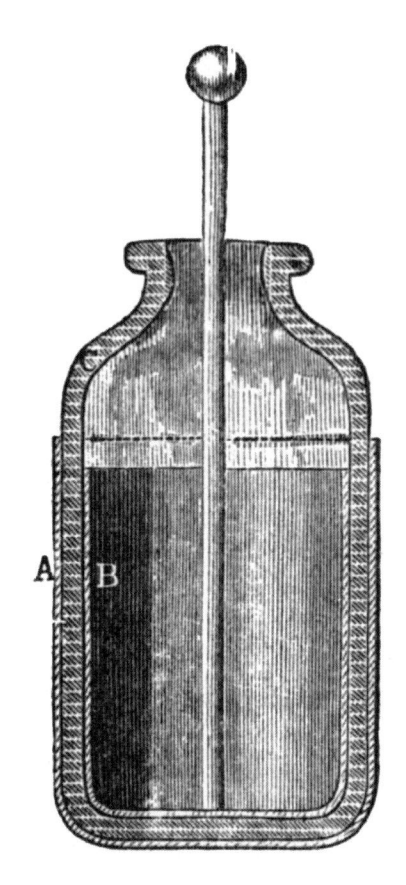

Galvani and Frogged Determination

The first modern development of the battery is credited to a 1749 discovery by Luigi Galvani, an Italian biologist who was studying how action is propagated in a living organism. Galvani's lab had a Leyden jar and an electrostatic machine, which Galvani would use to send charges into the muscles of frog legs, which he found would produce movement. Often this was done by touching the nerves of muscles with electrostatically charged medical utensils. History was changed when in one particular circumstance he had some long frog legs hanging from copper hooks in his lab, and as the story goes, he touched a region of nerve bundles with an *uncharged* metal instrument and caused the same twitching movement in the dismembered frog. Galvani realized that an electric charge was generated, and was stimulating the leg muscles of the frog, but in a way that did not require a charged utensil.

A Leyden jar, similar to those used by Franklin. These devices served to store electrostatic charge, and were common laboratory instruments in the late 1700s.

The chemical configuration that had been assembled by Galvani in this circumstance, between the two different metals (copper and iron) and the biological gel of the frog's fluids, are in general form the assemblage that we now know as the chemical battery. It is worth noting that the chemical process that produces the flow of electrical current that Galvani stumbled onto, as it was later discovered, is that which life utilizes for signal propagation throughout the body, but at a much smaller scale. In this sense, the battery really was a biochemical discovery, though not the discovery of some new distinct form of "animal electricity" as Galvani had thought. Also, as a historical cultural note, the widely popularized phenomenon of causing dead and dismembered animal parts to become animated by the application of electrical charge is said to be the inspiration

---

3. If Franklin had given the opposite names (calling resinous, rather than vitreous electricity *positive*), then we would not today have the odd circumstance where current is the flow of *positive* charge, while the electrons which actually do the flowing have *negative* charge.

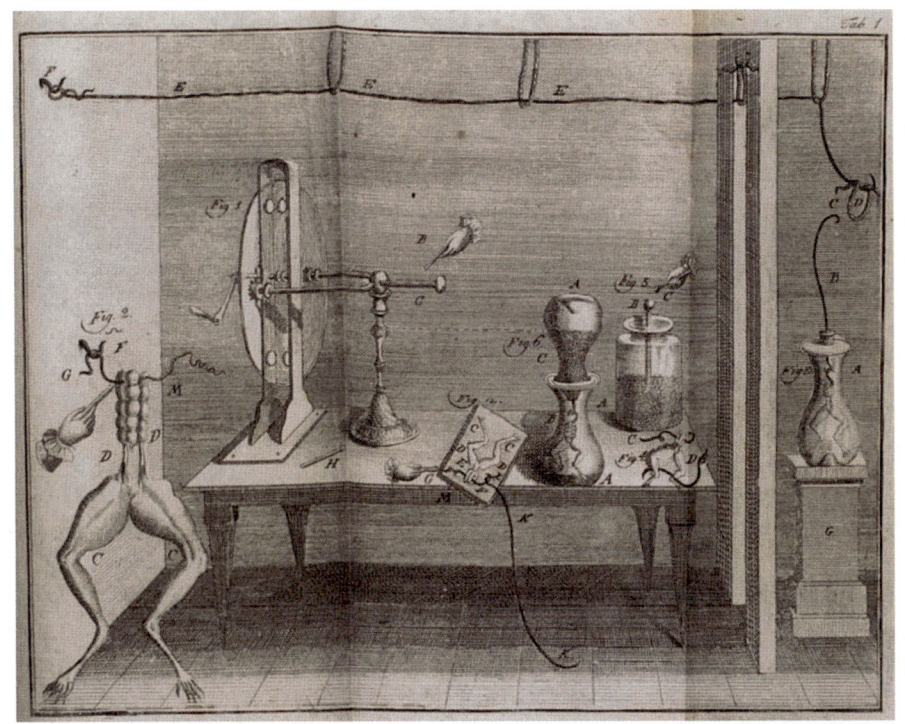

An engraving of Luigi Galvani's laboratory. Galvani's research into electricity and anatomy led to his serendipitous creation of the first electric cell.

for the Mary Shelley novel *Frankenstein*.[4]

In 1800, Alessandro Volta distilled and refined the process discovered by Galvani, reducing it down to a relationship between two different metals separated by an ionized solution, such as salt water, with the metals then connected to each other by a conductive material. This was found to produce the flow of electricity. Volta constructed what became known as the *voltaic pile*, a stack of alternating discs of copper and zinc, separated in turn by a piece of cloth saturated with salt water. It was found that when the briny cloth was removed from the pile, the amount of electric discharge was reduced to that of a single copper-zinc disc pair, whereas when the cloth was inserted within each pair in the stack, the charge was greatly magnified, increasing with each additional cloth-separated copper-zinc disc pair.

What we now understand is occurring is that the two metals, now called electrodes, are reacting chemically with the ionized solution. When atoms of the metal are induced to bond with atoms in the solution, which are more chemically attractive to them, they leave some of their electrons behind. The result is that the abandonment of the metal by the positively charged nuclei that go on to form new compounds with the solution, leaves behind negatively charged residual electrons, which build up to form an overall negatively charged state for that metal. A similar effect is occurring at the site of the other metal as well, the difference being that the developed charge is relatively positive in comparison with the first. This results in a potential difference between the two metals (voltage), which when connected to each other by a conductive material (one that allows for the easy flow of electric current), results in a flow of electrons from a place of high negative charge to one of less negative or relatively positive charge. The amount of flow is known as the current. It is like connecting a pipe from a mountaintop lake to a desert valley below and opening up the spigot.

It is again worth noting that this process is very chemical-specific, as you need the appropriate type of metals that will chemically react in a specific way with the proper solution to produce the necessary charge difference between them, which then will allow for the flow of electricity when the two are connected by a material whose specific atomic structure and macroscopic form allow for the relatively free flow of electrons. One of the earliest and most important experimental uses of electricity from batteries was the application of battery-supplied current to water to dissociate hydrogen and oxygen. Already in 1802, the chemist Humphry Davy was utilizing electrolysis to isolate a whole set of new metals from ores, including potassium, sodium, barium, strontium, and magnesium. Thus, the new science of electricity was opening up new potentials and insights in the older science of elemental chemistry.

Batteries continued to evolve over time as our understanding of the most efficient and effective elements to be used developed, but the fundamental concepts stayed the same. This includes rechargeable batteries, where the chemical reactions are induced to run in the opposite direction during the recharge phase, as for example, the lithium ion batteries that are at the heart of many of our most used electronic devices, such as computers, phones, and battery-powered cars.

It is extremely important to recognize that one of the first uses of this new power of controlled and sustained current was for the technology of the telegraph, though the concept remained mostly novel and conceptual until the next wave of electrical innovations and discoveries. With the telegraph, mankind now had the ability to carry

---

4. Mary Shelley's subtitle was "The Modern Prometheus." Considering the disasters that attend the use of science in the novel, what does this suggest about her view of the conflict between Prometheus and Zeus?

A voltaic pile of Alessandro Volta's design. This battery of electric cells (also known as elements) produced current by the interaction of the copper and zinc electrodes with brine-soaked spacers between the electrodes of each element.

out long-distance, nearly instantaneous communication, creating a new level of connectivity among the minds and thoughts of the human species. The first trans-Atlantic electronic communication was established in 1866, between Britain and the United States. Ten years later saw the patenting of the telephone by Alexander Graham Bell, and 25 years after that Guglielmo Marconi sent a wireless communication across the Atlantic, based on the discovery and experiments with electric waves made by Heinrich Hertz.

At the beginning of the 19th century, along with the parlor trick and novelty uses, the battery was the basis for conducting a whole new set of investigations and experiments into the nature of electricity, and eventually lead to the development of the theory of electromagnetism, where electricity and magnetism were unified under one banner, which led to the development of a great liberator of mankind, the electromagnetic motor.

Ampère: Beyond the Battery

With the ability to have sustained current, thanks to the development of the battery, André-Marie Ampère set up a group of experiments that looked at the relationships of different currents, flowing in different directions and configurations relative to each other. He showed that parallel currents moving in opposite directions repel each other while those moving in the same direction attract. Overlapping perpendicular current elements had no effect on each other at all. Ampère established the experimental and theoretical foundations for the development of the science of moving electric charge, which he called electrodynamics, even forecasting the existence of what we now know as the electron and deriving the relationship of electricity to the speed of light. To this day, one of the principal discoveries of Ampère, that of the "angular force," is a point of controversy, and has to a large extent been left out of textbooks, though some prominent scientists think it may hold the key to a more thorough understanding of atomic and subatomic processes.[5] A key to Ampère's work was his introduction to a principle discovered by Hans Christian Oersted, of the relationship between electricity and magnetism.

In 1820, Oersted observed that a compass needle, which responds to magnets and the magnetic field of the Earth, would move when brought close to an electric current. Moving electricity produced a magnetic effect. The inverse property of inducing current in a conductive material by the relative motion of a conductor and a magnetic field, was developed by Michael Faraday in the early 1830s.

This relationship between electric current and magnetism became, through the motor, the basis for the second industrial revolution, as the electric generator became the primary technology at all sides of the productive process. This is a result of the fact that magnetic and electric fields exist with a perpendicular relationship to each other, and the change of one produces the other. Take, for example, the generation of electricity at a dam with a hydroelectric generator which rotates a magnetic field through a bundle of conductive wire, which then transmits electricity across conductive lines where it is then transformed back into some form of mechanical action in an inverse way, as with a motor in an electric drill.

The electromagnet, which has replaced permanent magnets at the core of most generators and motors, has its roots in Ampère's work: creating a magnetic field by

---

5. Ampère's angular force is a result of his longitudinal force, whereby two collinear, opposite current elements attract each other and those moving in the same direction repel. This effect is denied and left out of most modern electrodynamics. For more, see "The Atomic Science Textbooks Don't Teach" in *21st Century Science and Technology*, Fall 1996.

running current through a circularly wound wire. He found that the strength of the magnetic field was directly proportional to the current and the number of windings of wire. The first generators were put into use by the mid-1830s, with the largest alternating current (AC) power generator of its era set up at Niagara Falls in 1895, producing a great deal of electricity from its hydroelectric generators. Hydroelectric power generation and transmission was fully exploited as a civilization-changing application under the guidance of the Franklin Roosevelt Presidency through the combination of a massive hydroelectric buildup and a rural electrification initiative, which saw electricity lighting up homes and chicken coops and powering the water pumps of farms across the country.

No longer was electricity a local urban resource, but with projects such as the Grand Coulee Dam and the Tennessee Valley Authority generating abundant amounts of new electricity, the vision of the Rural Electrification Administration could be realized. Farmers and other rural residents alike were now given access to the benefits of electricity which lead to not only an increase in the productivity of the farm, but also to cultural changes such as increased literacy and educational opportunity. Through the power of electricity, more people were liberated from the demands of farming as productivity increased, and could pursue other productive activities, further increasing the dynamic complexity of society. The United States as a whole was now operating in the field of electromagnetic potential, and was equipped with the power to not only stop global fascism, but to extend the field of progress on a global scale. Unfortunately, the untimely death of Roosevelt brought that momentum to a grinding halt, and in the main, that potential has been placed under the yoke of an oligarchical agenda.

The Motor of Human Progress

The first commercially developed electromagnetic motor was patented in 1837, though the practical and economically sound use of the motor did not take hold until the 1870s and '80s. From that point the motor became the principal technology for modern manufacturing processes, moving from the machine tool sector to the household consumer, from precision lathes, to conveyor belts, to the vacuum cleaner. The electric motor also found early use in transportation with the first electric trolley cars going into action in the late 1880s. With the

Julien Lemaître

Statue of André-Marie Ampère at the Museum of Electricity—the Maison d'Ampère—in Poleymieux, France. Ampère introduced the first detailed understanding of electrodynamics, and began to unify the principles of electricity and magnetism.

electric motor electricity could now be transformed into mechanical action, and as the electrical transmission grid spread, so did the productive potential. This was equally true for areas that already had manufacturing, where machines were all run by a complex of pulleys and axles connected to belts which were in turn connected to other pulleys and a larger gear system, which was all powered by an on-site steam engine.

Now, with electrification, all that was needed was a power line running in from a distant area, and individual machines could be much more easily plugged in, turned on, moved around, and changed out. The move away from belt-driven and on-site steam-powered production towards electromotive action began in the early 1900s. No place benefitted more from the use of wire-pumped electricity than deep mines, where combustion would be noxious and potentially dangerous, and pumped steam power was cumbersome and often impractical. Wires could now be snaked to otherwise hard to access areas. Ultimately this meant an increased density of machine power per capita as electrically powered motorized machines expanded and played an increasing role in more facets of the economy, from sawmills to blenders. Today, electricity is again being applied more widely to mass transportation, with electric cars and high-speed trains, where, depending on the case, the energy is either derived from an onboard chemical battery, or is drawn off the electrical grid. To this day, when it comes to the electrical grid, the source of the electricity is a mechanical electromagnetic generator, driven by either one, or a combination of gravitational, chemical combustion, or thermal exchange forms of work transfer.[6] Still in the design phase is the use of what is called "direct conversion" where the motion of charged particles is converted directly into voltage.

The Grand Coulee Dam, one of FDR's "Four Corners" projects. Grand Coulee's cheap and abundant electricity made the Pacific Northwest a center for aluminum and aerospace production, critical for winning World War II.

It's Not All About Machines

Besides the commercial use of the motor, electricity was exploited for its resistive interaction with certain materials, where the action of the retarded flow of the electrons is expressed as thermal and light energy. This resistive property of some materials is used in everything from a light bulb, where in many cases a resistive tungsten filament radiates energy in the form of bright white light, to an electric coil top stove where the primary amount of energy released is in the thermal range.[7] By the 1880s, light bulbs were gaining widespread use, thanks to the labors of such inventors as Thomas Edison and Joseph Swan, who made successive breakthroughs in understanding what materials would most effectively radiate light, as well as creating the proper vacuum conditions for the bulbs, which was necessary so that the filaments would not catch fire due to the presence of the highly combustive gas, oxygen. The lives of an increasing number of the world's people were transformed by the ability to have nighttime illumination on a grand scale, extending the productive hours in a day, whether it be at the farmhouse or the central city library.

Resistive materials, as, for example, the filament in a light bulb, have a somewhat inverse character to conductive materials. The theory is, that while conductive materials, such as the metals copper and gold, have a crystal lattice arrangement of their atoms which allows for a relative free drift of electrons, silicon-based resistors, on the other hand, have an impeding effect on the flow of electrons, creating a kind of friction to the flow. This type of friction energy is then radiated out from the surface of the material often as light or heat. Some materials, such as quartz crystals, translate the resistive action into vibrational action, which can be tuned and amplified by the amount of current and voltage that runs through it. This is the concept behind the quartz radio and watch.

Vibrating action is another key element in the technologies that utilize electricity, particularly ones that relay and transmit audio. By having a modulated oscillation of a magnet, a modulated flow of electricity can be induced, which can in turn be transformed into radio or other electromagnetic signals which can be captured and re-radiated as sound, effectively following the same steps back.

---

6. Excepting the insignificant (and foolishly wasteful) use of solar power.

7. The light produced by a regular tungsten incandescent light bulb is only a few percent of the energy it consumes.

The Power of Regulation

The property of resistance now plays a very critical regulatory role in all electronics, where resistors restrict the flow of current, in various degrees, to different elements of a circuit. This allows for the use of a variety of different components, which have varying degrees of sensitivity to current amount, to all be used in the same circuit. Without resistors, many of the components would require their own individual power source specified to their voltage and current requirements. Here again, the particular chemical composition and arrangement play a determinant role in how the properties of electricity will be expressed. For example, the resistors that are used in modern electronic circuits are often made of carbon mixed with some sort of ceramic, but as resistors these are best when used at small scales.

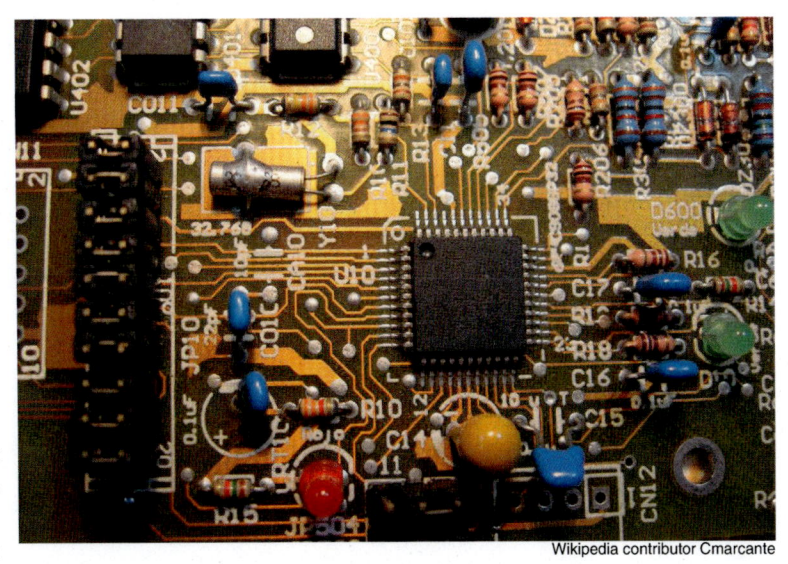

The semiconductor is an excellent regulator of voltage and of current moving through electrical devices. These are most commonly made of silicon or germanium, and doped with trace amounts of other elements to produce materials with a slightly positive or negative disposition. This doping transforms an otherwise non-conductive material into one that can now facilitate a flow of current by changing the electron configuration of the material. Semiconductors are in effect turned on and off in a process known as gating; the application of an electric field turns the material into a channel for the passage of current. Transistors are a type of semiconductive component that act as incredibly small electrical switches, replacing, in many cases, the mechanical relays and electromagnetic vacuum tubes used for this purpose earlier. This is crucial for being able to change and regulate the amount of current and voltage that enters the various components of an electronic device, allowing for increasingly more complex and diverse operations.

A printed circuit board, showing integrated semiconductor circuits (e.g., the central black square), capacitors (labeled with Cs), resistors (labeled with Rs), and other components. Semiconductor technology allows control systems to be orders of magnitude smaller.

Semiconductor integrated circuits are the basis of those electronics that are so dominant in today's world, where through the use of computers, we have created a much more refined use of electricity for the storage, transmission, and utilization of information, for communication and control of machinery. For example, the use of computers to mediate the control of airplanes and machine tools, translates the intentions of the programmer and user into complex and complicated precision actions. Computers express a remarkable utilization of the essential properties of moving charge, i.e., voltage, current, and resistance, and the specific material relationships to those properties. They represent, when applied properly, a significant extension of the physical power of man.

With the discovery and application of the principles of electromagnetism, mankind has tapped a fundamental dimension of the universe and opened up a new power. With all moves from a lower to a higher platform, society takes on a more multiply-connected nature, in terms of the number of interdependent connections that make up an individual's environment. Each individual has more power to transform the universe, but that power is a function of, and dependent on, an ever increasing array of technologies, institutions, and people. This is the natural tendency of mankind, to increase his power to act in the universe, through the increase in density of applicable fire, as a function of an increased knowledge of physical chemistry.

A CNC machine tool. Such automated devices can automatically change cutting tools and reposition work objects, allowing for high automation of part-machining.

The Nuclear Era
Man Controls the Atom

by Liona Fan-Chiang

Reach for the stars, and you may find one right here on Earth. If you do find one, it will have been created by man.

The nuclear age began more than a century ago, yet it still hovers in an adolescent stage. The prospect of full control of the atom, both by having taken control of splitting up or splintering large atoms (fission), and by the intentional merging of small atoms (fusion), was born not long after. After more than a century, one would have expected a society which had already graduated from performing small-scale experimentation on, and application of, matter and energy conversion, to having full reign over the natural and artificial transformation of material, the associated electromagnetic effects, and much more, with more precision than that expressed by the sun and life. Yet, something unnatural intervened. Anti-human policies, including know-nothing, anti-radiation environmentalism, exercised political and cultural control to stunt the advance of human evolution.

Here, we tell the short story of how the nuclear age was born, from investigations of naturally occurring phenomena, to the human attempt to master those processes and recreate them ourselves, in order to perhaps make clear where we are today and what it means to create states of matter and energy entirely natural, but yet entirely new.

Natural Transformations

Energy from Matter

Our experience with the constant flux of all material began with the discovery of radiation in the late 1800s and early 1900s.

Wilhelm Röntgen noticed in 1895 that when electricity was passed through a tube from which almost all the gas had been evacuated, a green glow appeared at one end of the tube. More astonishing was that an accompanying light, invisible to the eye but visible to his photographic plate, was detected. This light was so penetrating that even when passed through black paper or thin metal, it could be detected by his photographic plates.

Henri Becquerel, excited by Röntgen's discovery, began investigating. He had been studying fluorescent materials[1] intensely and thought this glow might be related to the mysterious fluorescing property of some materials. Choosing uranium potassium sulfate, a fluorescent substance, he indeed detected Röntgen's penetrating rays.

Events turned, however, when on a cloudy day, Becquerel's sample no longer fluoresced. Thought to be use-

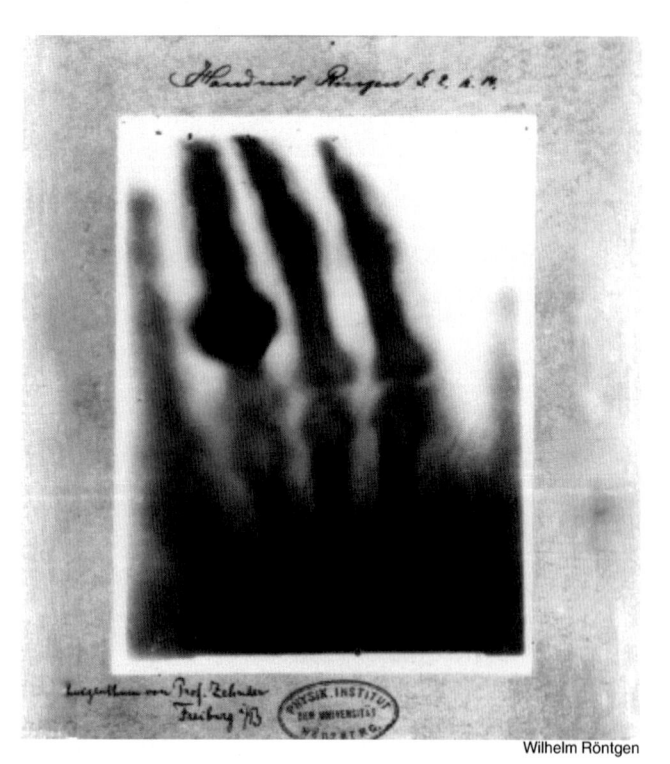

Wilhelm Röntgen

Wilhelm Röntgen's first X-ray image, his wife's hand, December 22, 1895.

---

1. Fluorescent materials emit light after exposure to sunlight. It was found that they also emit light after exposure to radium. In WWI, fluorescent material excited by radium was used for gun sights in order to aim in the dark.

less for his studies, he tossed the sample, photographic plate, and black paper in a drawer for several days. When he came back to it, discouraged, he developed the plate out of curiosity. Surprisingly, though the sample received no light and had not fluoresced, the clear imprint of the sample appeared brightly on the photographic plate.

Becquerel's rays, unlike Röntgen's, had not been stimulated: they seemed to emanate ceaselessly from the rock with no input at all. How could this be? Where was the energy coming from? Did conservation of energy cease to be a law?

To see what kind of force might have been at play, the uranium salt was subjected to the most demanding obstacle course: extreme pressures, extreme heats, magnetic fields, strong chemical reagents—nothing seemed to affect the rapidity of this constant efflux.

Marie Curie tested a tremendous number of substances, searching for other materials that emit Becquerel rays. She found that thorium salts also emit these Becquerel rays. The more salt, the more emanation. Becquerel's earlier discovery that the air around the emanation becomes a conductor of electricity greatly aided this search.

Marie Curie and her husband Pierre also isolated two other previously unknown elements, polonium and radium, the latter of which radiates one million times more than uranium.

Although the rate of radiation being emitted seemed not to be influenced by any force, Curie did notice that under strong magnetic fields, the rays, once emitted from the sample, could be affected, slightly deflecting and separating out into two beams: one that deflected in a strong magnetic field, and one that did not. When Ernest Rutherford applied an even stronger field, the undeflected beam again split into two: resulting in a beam that was undeflected and one that curved the opposite direction with respect to the first.

The characteristics of the first deflected beam resembled electrons in mass and charge, while the other deflected beam seemed to be a much heavier and oppositely charged matter. This posed another challenge. Was matter being continually emitted as well? Was matter being born? Where were these radiations being produced?

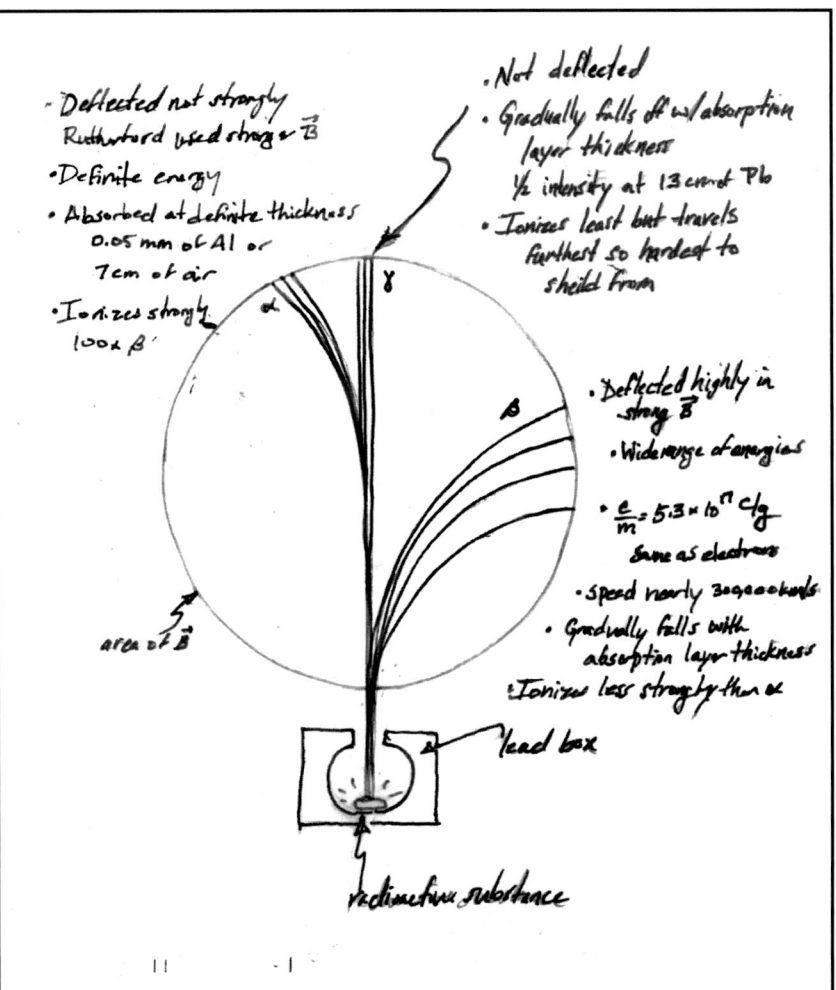

With a strong magnetic field, radiation was able to be split into three different rays with very different properties. They received their names, alpha, beta and gamma, from their different penetration depths. Alpha particles are charged opposite to beta particles, while gamma rays are not charged at all. Alpha particles came out with the same energy, while the energy of beta particles varied widely.

An Element Born

Röntgen made a bold hypothesis that part of the radiation was the helium ion, a hypothesis that was reinforced when helium was found in radioactive mines and not in others. This rule was so durable that the existence of helium was used to detect radioactive materials.[2]

---

2. Rutherford and Royds came up with an experiment to test this hypothesis. They placed a sample of radium inside of a tube which was thin enough to let alpha particles, the less deflected radiation, through. This tube was then placed in another tube with electrodes at each end and from which all the air was evacuated. When voltage was first applied to the electrodes, since there was no gas inside of the tube to conduct electricity, no current flowed. Two days later, current was able to flow in the larger tube, indicating now the presence of some gas. The characteristic glow of the gas betrayed its nature to be helium, confirming the hypothesis that alpha particles were actually the helium

In addition, already in 1899 the Curies had noticed that the air around thorium and radium was also radioactive. To confirm this observation, Rutherford and Soddy passed this radium air by fluorescent willemite.[3] The willemite glowed until the radium air was let out, at which point fluorescence stopped.

When current was passed through this radium air, the air glowed not only the characteristic colors of oxygen and nitrogen, but also of helium.

Most remarkably, during the course of their experiments, the radium air gradually lost its radioactivity. This was the first time that radioactivity from a radioactive material had been observed to decrease, and even drop below detectable levels. Experiments on other materials confirmed that they too had finite emission rates, but decreased much more gradually.[4]

The question still remained: If helium was being emitted from radium, what was becoming of the radium? Could it be staying the same, which would mean that helium was born from nowhere, or was the helium being fragmented off, thus leaving an element two steps down on the periodic table from radium? This product would be element 86, an as-yet undiscovered element whose existence was forecasted by Mendeleev.

When the radioactive radium emanation was carefully weighed by Ramsay and Gray (1910), they found an atomic weight of 222 atomic units, never before observed. Its chemical inertness placed the element in the column of the periodic table of elements occupied by xenon, placing the emanation squarely at element 86. It was subsequently named radon, to signify that it is a product of radium and has chemical properties of noble gases such as xenon, neon, and argon.

This resolved the challenge of the possibility of infinite quantities of mass. Matter was not just being emitted from a fixed material. Rather, radioactivity results from a process of a transformation of matter. Moreover, the process of transformation, its rate and type, was unique to each type of material, indicating that this transformative property is as much, if not more, an intrinsic property of matter as mass or chemical potential.

Artificial Transformations

In 1919, Rutherford accomplished and observed the first artificial disintegration of atomic nuclei. After many scattering experiments, in which he bombarded many different materials with alpha particles, he postulated that atoms have solid nuclei. He even arrived at a general formula that related the scattering angle of alpha particles to the charge of the target nucleus. Among these experiments, however, he found that when the target was a lighter element, his generalized equation did not work, in such a way that he was led to hypothesize that perhaps the projectiles were coming too close to the nucleus. He further hypothesized that this circumstance might potentially be used to splinter the nucleus.

He used the fastest bullets around: alpha particles traveling at 19,200 km/s emitted from what was then known as Radium C,[5] a product of radium. Several elements disintegrated into lighter elements when bombarded, but none heavier than potassium. Some elements lighter than potassium such as helium, lithium, beryllium, carbon, and oxygen also did not budge.

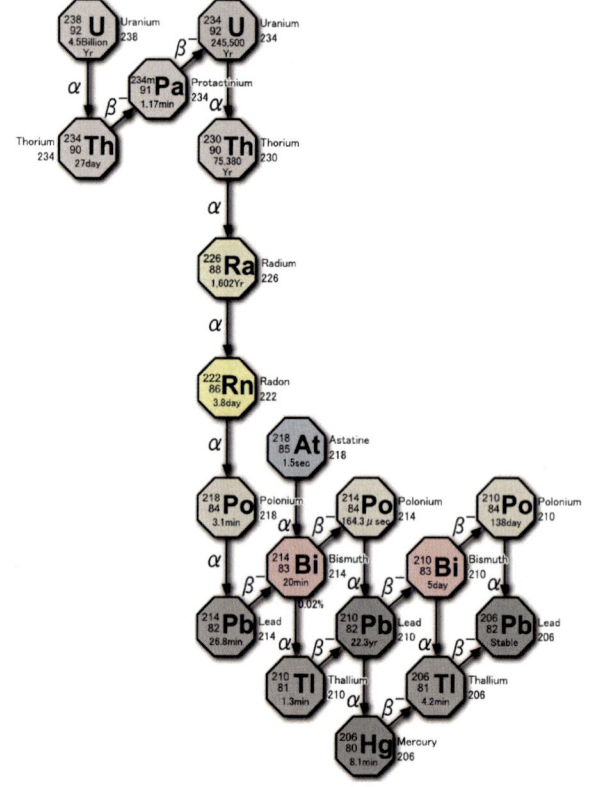

Wikipedia user Tosaka

Nuclei have unique characteristic transformation properties. Each has its own unique transformation rate, and its own way of transforming. The transformations form families of related nuclei, one of which is represented here. Even after people had discovered that thorium-D (ThD), a decay product of thorium-232, was just another isotope of lead-208, it was still called ThD. Why is that? As a product of thorium, it has more kinship with thorium than its chemically identical cousin, lead-207.

---

nucleus.

3. Like the sun, radium rays can excite fluorescent materials. Microscopic amounts of radium make screens of zinc sulfide, barium platinocyanide, willemite etc., glow in the dark.

4. Since the decrease was found to be asymptotic, radioactive lifespans are measured in half-lives, the time it takes for half the sample to decay, and therefore only half the emission rate to be measured.

5. Bismuth-214.

Enrico Fermi also began transforming materials artificially. He took the heaviest known element, uranium, and attempted to create a new, even heavier element. He got a surprise. When he bombarded uranium with neutron radiation, the sample began to exhibit a new, complex radiation consisting of beta rays. It took Fermi over five years to disentangle the complex radiation data, and he found at least four different decay rates

Fermi found three isotopes of uranium, that is, three different weights of the same element uranium. These isotopes had exactly the same chemical properties, differing by weight, and, as was to be discovered, by nuclear transformation properties. Half lives of 10 seconds, 40 seconds, 13 minutes, and 90 minutes were measured. Fermi expected that at least one of these products must be element 93, the next higher element after uranium. None of the elements from 86-92 had half lives in the range of 13-90 minutes, so Fermi assumed that the products must be elements with atomic number of 93 or above.

Lise Meitner, Otto Hahn, and Fritz Strassman were convinced the problem was more complicated. The curve of intensity of attenuation of radiation was different at different times, indicating that other radioactive substances were arising sometime *after* irradiation. Their experiments showed nine half lives, with the highest element being element 97. After meticulous separation, they found three parallel series of transformation.

Something else was also occurring. Some of the radiation products were chemically indistinguishable from lanthanum (element 57) and barium (56). These were elements much too small to have been a disintegrated product of uranium (92), a process which was assumed to end at lead (element 82). Or so they thought.

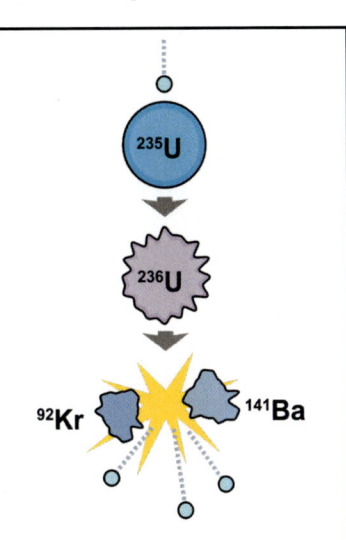

When uranium-235 is bombarded by a neutron, it fissions, or splits apart. Besides its fission fragments, it also releases neutrons, stimulating other uranium atoms to fission, etc., resulting eventually in a self-sustained chain reaction, and a continuous energy source, until so much of the fuel has been converted that the reaction is no longer self-sustaining.

Lise Meitner alone came up with the bold hypothesis that these products were in fact not results of simple addition and subtraction, such as in a decay or bombardment process, but of *division*. She hypothesized that if the atom had actually split nearly in half, then there should be two corresponding elements which belong in the region of the periodic table at nearly half the atomic number of uranium, and that then there should be a corresponding energy release equivalent to the mass lost that can be calculated from Einstein's $E=mc^2$.

U-236   Mass = 236.0455 u
Ba-141   Mass = 140.9144 u
Kr-92   Mass = 91.9261 u
3 neutrons   Mass = 1.0087 u

Mass Defect = U-236 − Ba-141 − Kr-92 − 3 neutrons
= 0.1790 u

$E = mc^2$ = extra energy released
= 0.1790 u × 1.66×10^{-27} kg/u × (2.998×10^8 m/s)^2
= 2.67×10^{-11} J extra energy

If fission had in fact occurred, if her hypothesis was correct, a specific amount of extra energy should be detected in the reaction. This was indeed confirmed.

Other fission products were subsequently found, including technetium (43), ruthenium (44), and rhodium (45). Over 100 papers on fission were published in the following year. Afterward, many others, from germanium (35) to samarium (62) were found in the fission products.

A whole new domain of the elements became potentially open to control by society. New isotopes and new elements were found among the fragments. In addition, the idea to harness the relatively large amounts of energy released by the atom soon became obvious and inevitable.

By looking at the mass excess of each element, it was hypothesized that the fusing together of the nuclei of very light atoms (such as hydrogen) would produce new atoms that would weigh less than the total weight of the original constituents, thus releasing energy in the process. By using the lightest ones, such as hydrogen, helium, etc., as in deuterium-helium-3 fusion, the energy released could be an order of magnitude more than from fission.

Particle Accelerators

In order to have more control and more power to transform the atom, artificially energized particles, much more energetic than those available naturally, were sought after.

The first successful attempt to transform nuclei by accelerating protons was by Cockcroft and Walton. Since

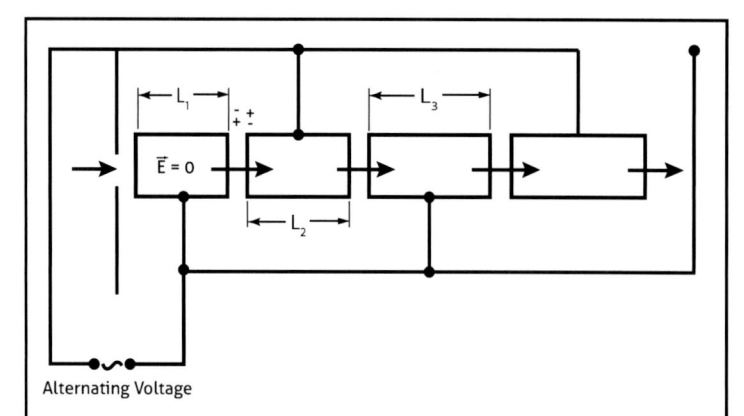

A linear particle accelerator which uses alternating current. Inside of the tubes there is no electric field so the particle does not accelerate. Once it exits the tubes, the particle experiences an electric potential which accelerates it across the gap into the next tube.

the proton is charged, it can be influenced and accelerated by a voltage difference. By placing several voltage differences in succession, they were able to accelerate a particle to 800,000eV.[6] At this potential, a proton would be accelerated to about 12,000 km/s.

This room-sized apparatus, however, reached its limit. The next step was advanced by Lauritsen, who came up with a way to use AC voltage. Alternating current was much easier to supply than DC voltage, and this apparatus greatly improved particle acceleration.

Terminals are connected to the AC voltage supply. The space inside the tubes is electric field free: inside, the particle does not accelerate, and travels at whatever speed it entered.

The difficulty this apparatus introduced was that as the particle accelerated, the length of tube it travelled before the voltage shifted increased, so that each successive tube had to be longer $L_1:L_2:L_3:L_4:L_n=1:\sqrt{2}:\sqrt{3}:\sqrt{4}:\sqrt{n}$, eventually requiring enormous sizes.

The next step would be to introduce a magnetic field. By introducing a magnetic field, the particle can be made to move in a circle, around the axis of the field. The key to this design was the fact that given a constant magnetic field, and a constant mass of particle, the period of rotation remained constant no matter what speed. In other words, as the particle was accelerated by two electrodes placed on either side of a short gap that cuts from the center of the circular apparatus to its circumference, the particle's path would increase its radius, but maintain the same period. Thus, if the frequency of the alternating electric field were set to the corresponding period, the particle could be accelerated continuously, limited only by the radius of the apparatus.

Further challenges were posed by this. Remember that the period stays the same as long as the mass and magnetic field stay the same. However, as a particle accelerates to velocities on the order of the speed of light, this is no longer the case. The particle's mass begins to change, and so does its period of rotation, resulting in eventual desynchronization with the frequency of the alternating field and thus eventually deceleration.

Particle accelerators have continued to improve to this day, resolving this issue and introducing others which result from conditions never before created by man. Humans now create artificial transmutations with artificially accelerated particles.

This ability has expanded the periodic table of elements to over 3,000 isotopes, each with their unique nuclear transformation properties.

Among these are new elements, unique to the nuclear age. For instance, for a long time, besides the transuranic elements, sub-uranic elements 43, 61, 85 and 87 were unknown. Even to this day no stable isotopes of these are found (technetium, promethium, astatine,[7] and

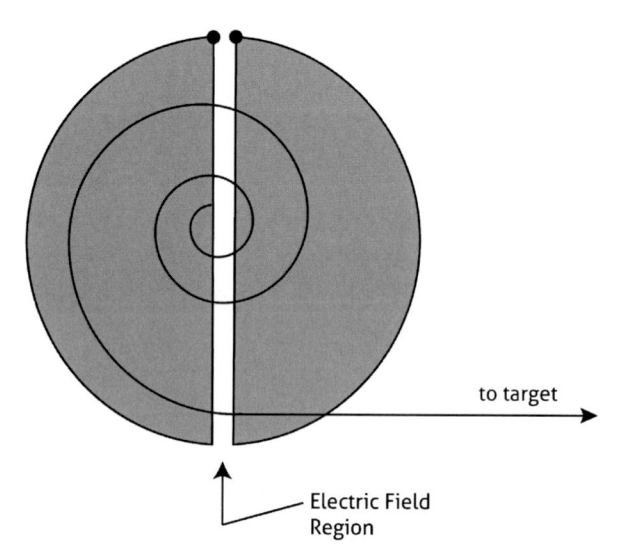

In a cyclotron, particles are injected into the device in the center of the apparatus, after which they orbit about the magnetic field, which in this case is directed perpendicular to the page. As long as the mass of the particle and the strength of the magnetic field remain constant, the orbit of the particle will only increase in size, while the period will remain constant. An electric field is generated in the gap, fluctuating at a rate that coincides with the orbital period of the particles. This electric field accelerates the particle every time it crosses the gap.

---

6. eV refers to a unit of energy called an electron-volt. It is the amount of energy gained or lost when an electron traverses an electric potential difference of one volt. It is equivalent to 1.602×10^{-19} joules.

7. From the Greek *astatos*, meaning *unstable*.

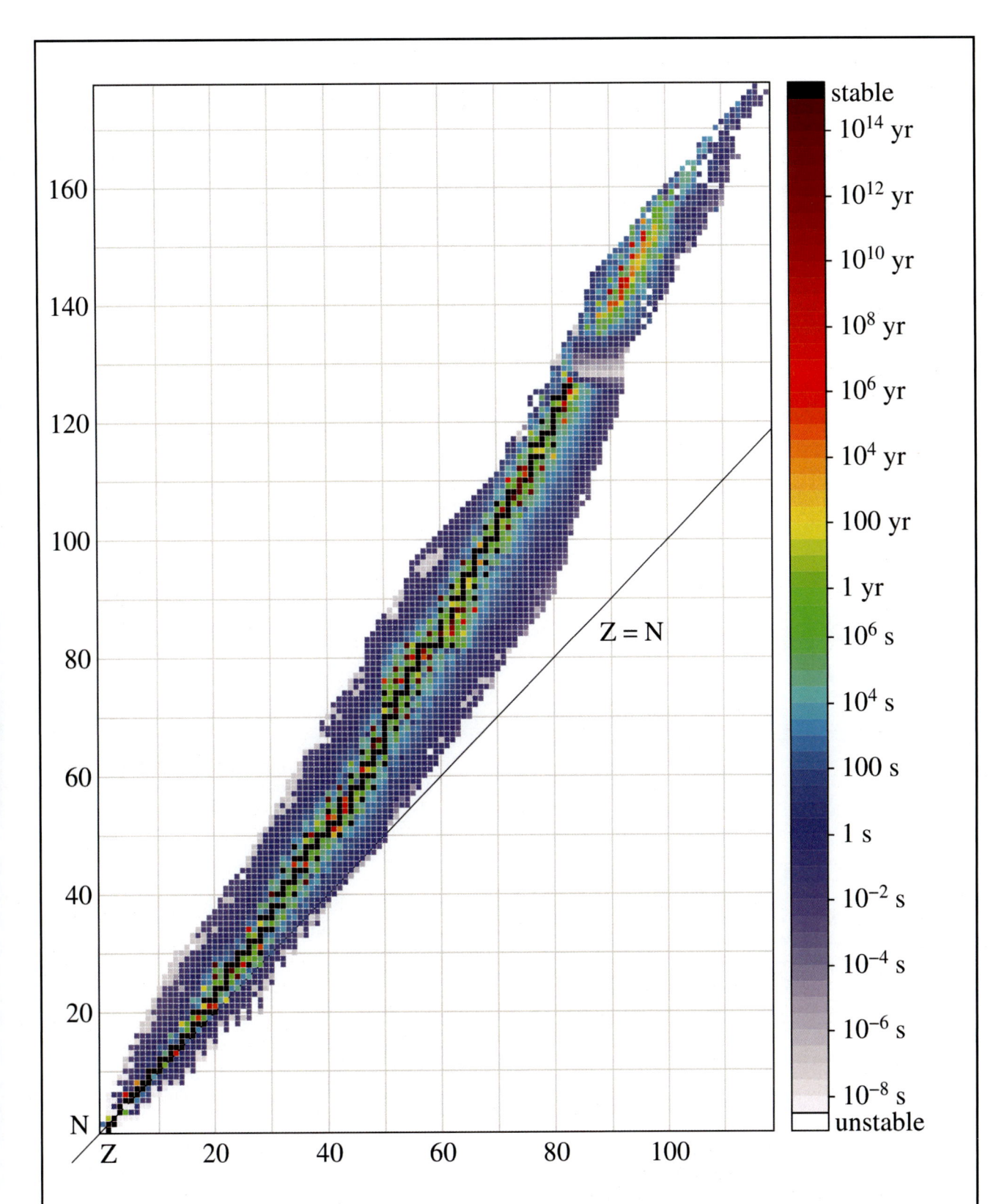

This table of nuclides lists all of the known nuclei. The x-axis represents the number of protons, while the y-axis represents the number of neutrons. Therefore, any column of nuclei represent the isotopes, or variations of the same element. The colors here signify the strength of radiation of each, by indicating the half-life of that particular nuclide, the amount of time it takes for half of any sample to decay. Black indicates stable nuclei.

The Continuing Gifts of Prometheus

francium). Only through artificial nuclear transformations have these elements existed in large enough quantities to determine some of their chemical properties.

Many of the radioactive isotopes play a crucial role in medical diagnosis and treatment. Some ten million medical diagnostic procedures annually use the artificially created technetium-99m, derived from molybdenum-99, a product of uranium fission.

Though now in the realm of possibility, we have not yet reached the point where we can freely traverse the now over 3,000 known nuclides, integrating the use of their various characteristics of transformation throughout the economy.

More advanced control over the process of fusion, in combination with our knowledge of the fission process, which still has much room for advance, will be required to fully realize this new potential.

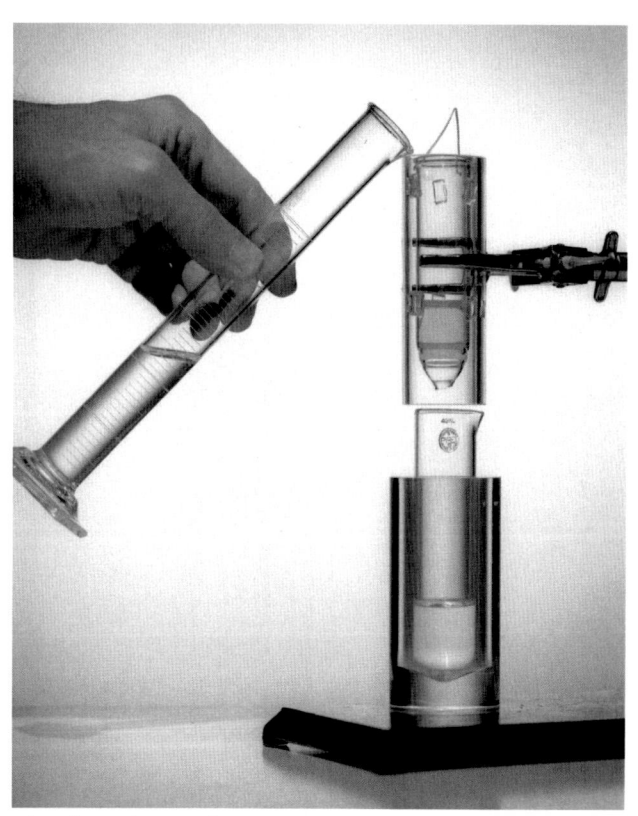

The first chemical separation of technetium-99m from molybdenum-99.

Conclusion

The launching of this great era now lies over one century ago. Yet, despite all of the advances of the 20th and now 21st century, we are far behind what we had set ourselves up for. We still cannot at this point say that the nuclear age has matured. Many fundamental questions have yet to be addressed, and many citizens educated about their heritage, before we can truly say that the nuclear age is here.

Still so many open questions are waiting. For example, what is the structure of the atom? William Draper Harkins[8] and Robert Moon posed unique hypotheses about the structure of nuclei, hypotheses which, if developed, could potentially enable us to order the various characteristic properties of nuclides, such as their unique decay rates and energies, and even forecast them. Their work is yet to be followed through to the point that a higher universal principle can serve to organize all nuclides, just as the periodic table, formed out of a higher organization, involving relations of action with respect to all others, rather than any isolated physical description, served to organize all elements, known and unknown.

Is there a whole domain of material and even principle which remains subdued by the noise of mixtures of isotopes? What new types of properties will become apparent once isotopically pure materials and compounds are fully explored and mass produced? There are already examples of diamonds made of pure carbon-12 or carbon-13 which are harder than regular diamonds, and of isotopically controlled silicon for computing.

What is the relationship between a full understanding of the nucleus and the study of life? Life is very sensitive to isotopic variations. It has isotopic preferences much like its preference for right-handed sugars and left-handed proteins. What is the role of nuclear transformations within the body? What can we learn about life were we to take full control over the nuclear domain? Or, rather, what can life teach us about the nature of the nucleus? There are already some very important medical applications in use. Much more needs to be done, and can be done once these degrees of freedom are explored and conquered.[9]

Uniqueness of isotopic ratios in astronomical data has shown, and will continue to show us unpredicted generating processes, invisible to the purely chemical domain.

Ultimately, the fundamental shift from a matured nuclear age to the next platform will be seen not in new technologies, but in how people identify themselves. Do most people think of themselves as conductors of stars? And as a species which lives among the great powers that control elements? Why not? And what would be the consequences if we did?

That is the story of Prometheus.

---

8. Harkins made many hypotheses based on experiments done on nuclear decay at the beginning of the 1900s. Based on alpha decay, he hypothesized that the structure of the nucleus might consist of units of helium and hydrogen nuclei.

9. Rouillard, Meghan, "Isotopes and Life, Considerations for Space Colonization," *EIR*, Vol. 37, No. 25, 2010.

Part 2

Physical Chemistry
The Promethean Future

NAWAPA and Continental Water Management
A Promethean Task

Helium-3
Stealing the Sun's Fire

Continental Water Management:
A Promethean Task

by Jason Ross

Among the great Promethean projects of our era is the planned and updated North American Water and Power Alliance (NAWAPA XXI), a proposed treaty agreement between the three nations of North America to jointly construct a system for the collection, regulation, and distribution of water. This continental system of water management would significantly increase overall biospheric and economic productivity by partially correcting the poor geography of the North American continent, and it would represent a decisive break from the anti-human environmentalist cult belief that any changes to "nature" made by man are inherently sinful.

Just as Franklin Roosevelt's physical-economic New Deal programs, such as the Tennessee Valley Authority, Rural Electrification Agency, and Grand Coulee Dam made possible higher levels of actual economic productivity (as opposed to the purely monetary financial activity of gambling, whether in Las Vegas or on Wall Street), NAWAPA XXI represents, today, a powerful vision: a rejection of the Wall Street and Green ideologies, and a specific, powerful project with massive economic dividends that will employ and improve the skills of millions of workers. Agricultural land will more than double in some states, overall living standards will increase, advances in industries for the project (nuclear power, tunnel boring, large-scale earth moving) will flow to all areas of the world's economy, and a currently largely unskilled generation of North Americans will have the opportunity to gain experience while participating in something worthwhile.

Had NAWAPA been built half a century ago when it was proposed in 1964, the extreme drought currently decimating agricultural potential in California would not be a problem, as the enormous water storage potential of the system's reservoirs would serve to even out wet and dry years. *Today, we are still the victims of nature; tomorrow, we will increasingly be its masters.*

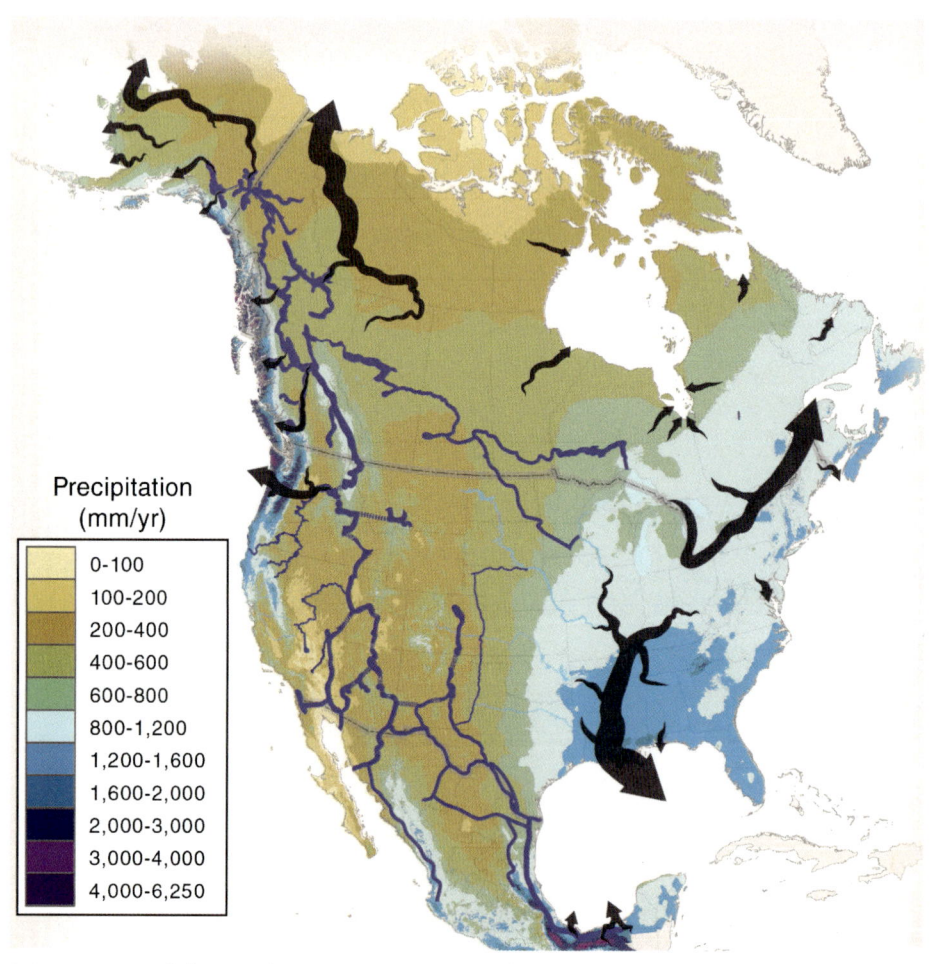

Water routes of the North American Water and Power Alliance, superimposed upon a map of rainfall patterns and river flow across the continent. The disparity of moisture levels leads to astonishing inefficiencies in the utilization of the continent's land and water.

The Productivity of Water

The value of any product in an economy depends on its context, and water is no exception. Due to the unfortunate geography of the North American continent, a great deal of water and land are wasted, falling far short of their potential to participate in both the broader biosphere and in human economic activity.

The three primary limiting factors for biospheric productivity (as measured by photosynthetic carbon incorporation, for example) are: sunlight, water, and temperature. The frigid temperatures and long, dark winters of the northern reaches of the continent severely limit the biospheric potential there, while the productivity of the Great American Desert, with plenty of plant-supporting sunlight and warm temperatures, is stymied by a lack of water. Dry land lies unused, and a large percentage of the water falling on the northern and western regions of the continent flows swiftly back to the ocean, unused by life on land.

Let us now quantify these qualitative characterizations. By combining rates of photosynthesis with studies of water flow, it is possible to determine the amount of plant growth per amount of water. This analysis reveals that every volume of runoff water in Alaska, Yukon, and British Columbia supports only *one-fifth* the photosynthesis of that same volume of water in the U.S. Southwest and northern Mexico. That is, each volume of water in the Southwest is currently five times more productive than the same quantity of water in Alaska.[1]

Now that we can compare *current* productivities of water already in these regions, what would be the effect of relocating water to different regions? To determine this, the relative importance of the three primary factors of sunlight, temperature, and water must be considered. Since water is not the limiting factor in the northern part of the continent, decreasing the amount of water there will have much less impact than increasing the amount of water in a dry region. That is, relocating the proposed amount of water from the northwest will have almost no impact on the biospheric productivity of that region, while it will have a dramatic impact on the dry regions it is brought to.

Under current conditions, each cubic kilometer (km^3) of runoff water in the northwest is associated with one million tons of carbon being incorporated into plant life, a figure known as net primary production (NPP). In the southwest, the figure is five million tons per km^3 of water. Assuming the water re-routed to the southwest will be just as biologically productive as current water, the over 100 km^3 of water brought by NAWAPA XXI can be expected to increase biological productivity by over 500 million tons per year, which would *double* the biospheric productivity of the region. Across the entire continent, NAWAPA XXI brings the potential for an increase of 10–15% in the productivity of the North American water cycle. It would be absolute foolishness not to take advantage of the opportunity to increase the productivity of the continent in this way.

Benjamin Deniston, based on figures from R.J. van der Ent et al., doi:10.1029/2010WR009127

The regrettable geography of North America. The arrows on this map indicate the flow of atmospheric moisture. Unlike our more fortunate neighbors in South America, where westward-flowing precipitation can move deep into the continent, the central mass of North America, lying farther from the equator, has eastward-flowing winds bringing water from the oceans, but this water does not get far into the western part of the continent, due to coastal mountain ranges.

---

1. Photosynthesis can be measured by the mass of carbon newly incorporated into living matter, in tons of carbon per square kilometer per year. This is referred to as "net primary productivity" (NPP). NASA provides values of NPP for the entire planet throughout the year. The northwest portion of the continent (Yukon, Mackenzie, Fraser, Columbia, and the northern half of the Pacific Seaboard) has an NPP to water ratio of 1 million tons of carbon per cubic kilometer of runoff water, per year. The Southwest basins (Great Basin, Colorado, Rio Grande, El Salado, and the corresponding southern region of the Pacific Seaboard) and High Plains Basins (Nelson, Arkansas / Red, Missouri, and the Texas Seaboard) have an average NPP ratio of about 5 million tons of carbon per cubic kilometer per year.

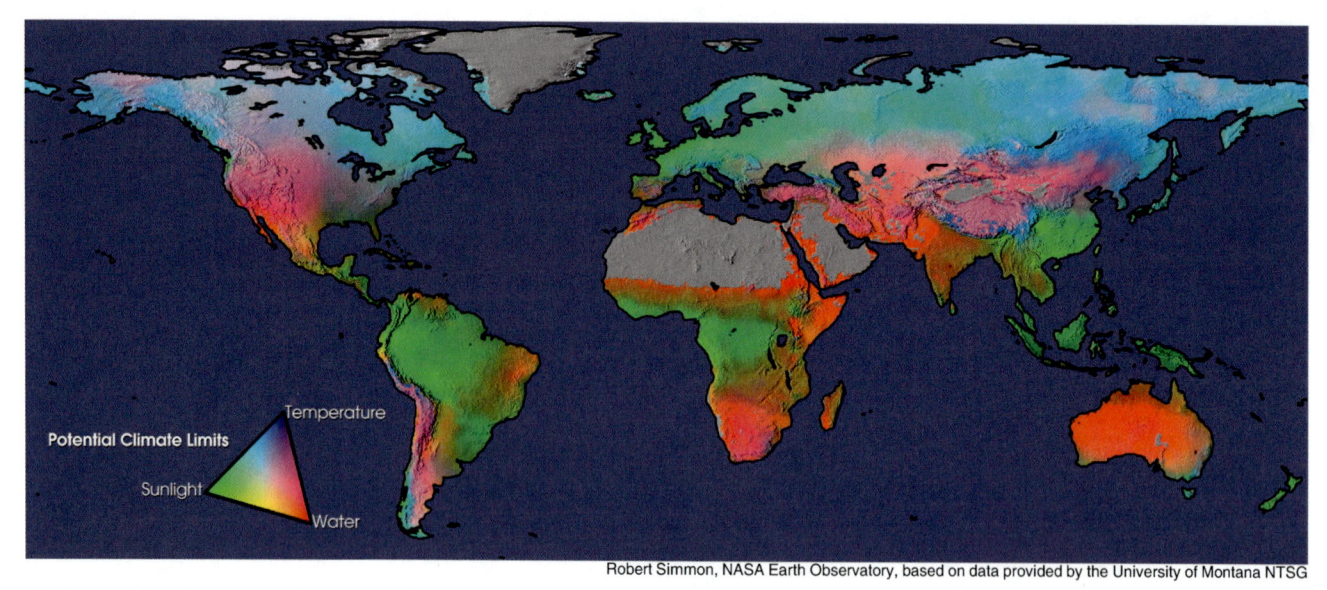

Limits to biospheric productivity. The colors on this map indicate the extent to which sunlight, temperature, or water limit the productivity of the land. By moving water from regions of plenty to regions of scarcity, life increases.

It is also necessary to take into account the role of plant life recycling water, by transpiration. To put this in context, note that every year, 40,000 km³ of water from the oceans falls as precipitation on land, while 73,000 km³ enter the atmosphere from land, and then fall again on land. This means that the water falling on land as rain, primarily came from land, not the oceans. Of this 73,000 km³, some comes from simple evaporation (from rivers, lakes, streams, and soil) while the rest comes from transpiration (evaporation from plants). Recent estimates, based on isotopic fractionation of water molecules, suggest that over 80% of the water entering the atmosphere from land comes from plants.[2] What will this mean for the continent as we wilfully redirect water flows and make deserts bloom? Every volume of water introduced directly by NAWAPA XXI will fall again as rain multiple times before making its way back to the oceans, its use being extended and multiplied by life.

All told, this means the transformation of the continent's climate and biosphere, far beyond adding irrigation water for cropland. The weather-moderating and moisture-enhancing effects of NAWAPA XXI will spread beyond the regions directly receiving the redirected water.

NAWAPA XXI System Overview

That said, how will this water distribution be brought about? First, some details are needed on the wide discrepancy of rainfall distribution on the western part of the continent, due to the particularities of the Pacific Ocean weather system. The area stretching from Alaska and Yukon down to Washington State has forty times the annual river runoff of the Southwest and northern Mexico. Through a system of man-made canals and utilization of helpful continental topographical characteristics, a 2,000-mile reservoir system can collect and distribute runoff in the most efficient manner possible. The design proposes incorporating roughly 20% of the runoff of each northern river into the system to be collected for distribution. And unlike the original 1960s NAWAPA proposal, which released a portion of the collected water through hydro plants to generate the electricity to pump the remainder through the mountains, the use of nuclear power means that all the water collected will be available for delivery.

The collections from the Susitna, Copper, Yukon, and Taku Rivers (see map) are pumped from 2,100 to 2,400 feet into the Stikine Reservoir, which receives water from the Liard Reservoir, before joining with the Nass and Skeena Reservoirs, themselves flowing into Babine Lake and Stuart Lake at 2,330 feet elevation. If 20% of each river's annual mean runoff is collected, approximately 87 million acre feet per year (MAFY) would flow out of Stuart Lake into a man-made canal. Of the 87 MAFY flowing out of Stuart Lake, some 70 MAFY will be pumped into the Rocky Mountain Trench Reservoir, while around 17 MAFY will be diverted into Lake Williston for the Prairie Canal, where it will join the 33 MAFY collected from the Mackenzie basin streams. In the Rocky Mountain Trench, 20 MAFY will be added from the upper reaches of the Fraser River, and 10 MAFY will be added from the upper Colum-

---

2. The ratios of oxygen-18 to oxygen-16 and hydrogen-2 to hydrogen-1 are different in water that has transpired from a plant and water that has simply evaporated. Thus, measuring isotope ratios in atmospheric moisture provides a means of estimating the relative contributions from plant transpiration and evaporation. See "Terrestrial Water Fluxes Dominated by Transpiration," Jasechko, et al., *Nature*, April 18, 2013, as cited in Benjamin Deniston, "The End of the Green Paradigm: Texas to California with NAWAPA XXI", in press.

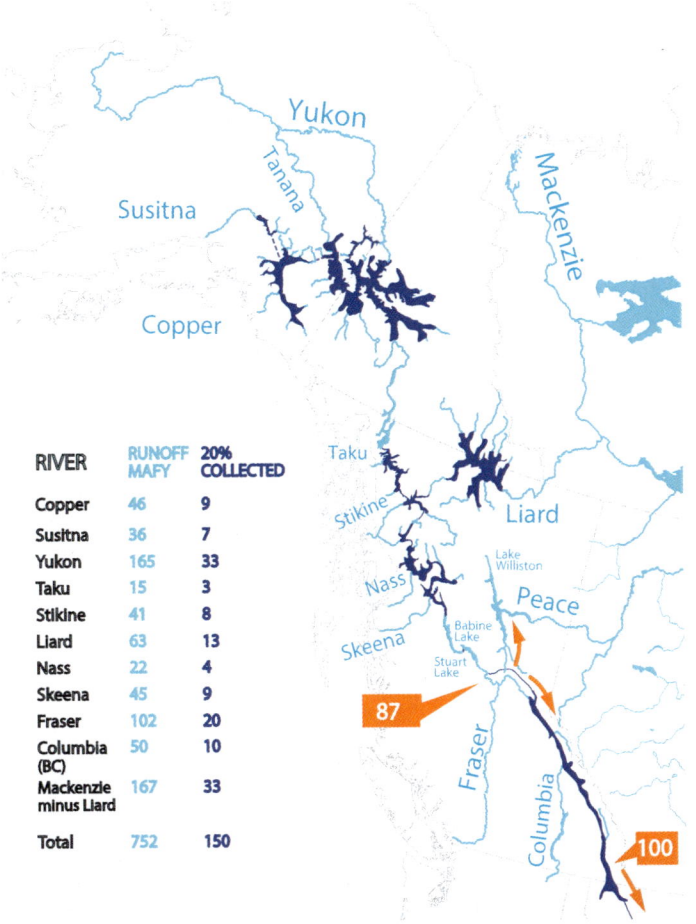

RIVER	RUNOFF MAFY	20% COLLECTED
Copper	46	9
Susitna	36	7
Yukon	165	33
Taku	15	3
Stikine	41	8
Liard	63	13
Nass	22	4
Skeena	45	9
Fraser	102	20
Columbia (BC)	50	10
Mackenzie minus Liard	167	33
Total	752	150

Map of the collection system of NAWAPA XXI, including annual runoff in millions of acre feet per year (MAFY), and the proposed collection volumes of 20% of that total.

bia. The 100 MAFY flowing out of the Rocky Mountain Trench will be pumped through the Sawtooth Lift in Idaho and diverted multiple ways throughout the Southwest and northern Mexico. Once the design phase is completed and construction begins, it is feasible to adopt an accelerated timetable and apply new technologies, to bring pieces of the system online only years after it begins, with the main trunk line completed in ten to fifteen years.

Once the completed NAWAPA XXI system is built, water will be able to be delivered to every major river system and region of the continent west and north of the Mississippi. All of the plans will form an interconnected grid across the continent which will be managed as a single system.[3]

Nuclear desalination facilities along the completed NAWAPA XXI irrigation systems will augment the effect of the canals by recycling water more quickly, as well as increasing the total amount of water available. The completed system will allow for wide-scale biospheric engineering and directed water recycling, creating a broader hydrological effect than the direct water contributions from the distribution system itself. Scientific institutions which study the effect of moisture in arid regions toward effecting changes in local climate and weather patterns, will collaborate in planning specific types of land cover for specific regions, and enacting other techniques of weather modification. Reservoirs will also be maintained to maximize aquaculture.

The Needed Mission

There can be no equivocation on pursuit of NAWAPA XXI. While the current drought condition underscores the long-term necessity of this project, the more fundamental fact remains: the natural course for the human species is to develop and implement new technologies to improve living standards and move towards the ultimately achievable goal of a society in which all people have the opportunity to contribute something of lasting, durable value with their lives. On the path towards securing a lasting, efficient, physical immortality for all, greening and improving the continent with NAWAPA is an obvious step.

---

3. This is a very brief overview of the project. Readers are encouraged to consult the *21st Century Science & Technology* Special Report "Nuclear NAWAPA XXI: Gateway to a Fusion Economy" for more details. Available at:
http://21stcenturysciencetech.com/Nuclear_NAWAPA.html

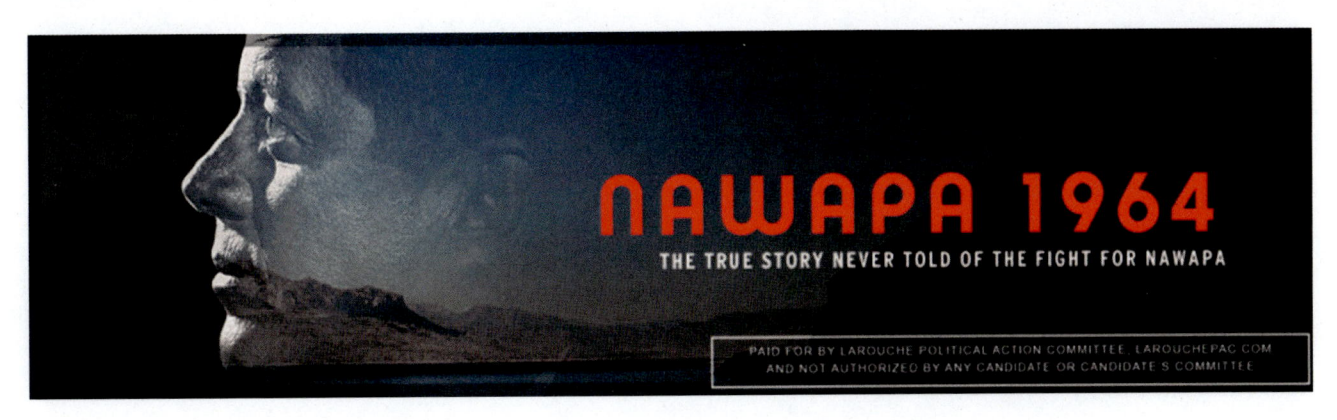

Helium-3
Stealing the Sun's Fire

by Natalie Lovegren

The evolution of the science of chemistry has enabled us to achieve control over energy and matter by finer and finer degrees of precision and with greater density of power. Each discovery has afforded us new dimensions of knowledge which allow us to extend our curious reach out into the bigness of space, and down into the vast minuteness of matter with greater and greater power on a smaller and smaller scale. The degree to which we can advance such a power over nature, and utilize these myriad "gifts of Prometheus" defines our existence as a species.

Here we unravel the case of one singular substance, and investigate the change of its identity, and of its economic value throughout the advancement of physical chemistry.

The Strange Case of Dexter Gas

In May 1903, residents of Dexter, Kansas, were thrust into fits of sheer jubilation after a newly drilled well started spewing forth natural gas at the rate of 9 million cubic feet per day before it could be capped. With the promise of cheap fuel and lucrative industries coming to town firmly in mind, the people sprang into action, planning to celebrate the discovery of this "howling gasser" with games, speeches, music and a lighting ceremony that promised residents "a great pillar of flame" that would "light the entire countryside for a day and a night." Yet when the time came to light the well, the gas refused to burn. Mystification and dejection ensued.

Word quickly spread across the state, piquing the interest of University of Kansas geology professor Erasmus Haworth, who brought samples of the curiously nonflammable "Dexter gas" back to Chemistry Hall at the University. There, two chemistry professors, Hamilton P. Cady and David F. McFarland, began two years of extensive research and analysis of the strange gas.[1]

Finding huge pockets of "free" natural gas to be burned for fuel was an exciting prospect at this time in the United States. But that wasn't always the case.

It had been known since antiquity that invisible flammable gases could come out of the earth. The infamous Temple of Apollo at Delphi was built upon a fissure in a rock, whence seeped a burning gas, because they believed the flame to have a divine source. The oracle who resided at the temple was said to be inspired by the flame, which enabled her to make prophesies on behalf of the god Apollo.[2]

But this gas merely fueled the superstitions—and decline—of the Greeks.

The development of natural gas for commercial economic purposes required the firm establishment of modern chemistry. It first required going beyond the mere *observation* of fire, and of gases burning, to understanding what burning was.

Dmitri Mendeleev, discoverer of the periodic table of elements, wrote in his brilliant work, *The Principles of*

"The Oracle of Delphi Entranced" by Heinrich Leutemann

---

1. John H. McCool, Department of History, University of Kansas kuhistory.com/articles/high-on-helium

2. Both Aeschylus and Plutarch (who was one of the priests of Apollo, responsible for interpreting the oracle) attributed the oracle's powers of "prophesy" to her inhalation of gases coming from the ground. Ethylene, a component of natural gas, is known to have hallucinogenic properties. A 2001 study, published by *Geology*, corroborates the claims of the ancients by detailing the intersection of two geological faults directly beneath the temple, as the source for such fissures in the rock which emitted these natural gases. Natural gas during the collapse of this once great civilization, was thus, not a resource, but a symbol of a nexus of usurious money lending, sophistry, and superstition, as evidenced by the willingness to consider the euphoric delusions of an intoxicated woman as sacred political wisdom. See also: Humphreys, Colin J. *The Miracles of Exodus*. London, 2003.
Papert, Antony. "Speaking of Delphi..." *EIR*, 21 October: 2005.

"An Experiment on a Bird in an Air Pump" by Joseph Wright of Derby, 1768.

Chemistry, that one of the reasons for the tardy progress of chemical knowledge was the pivotal importance of invisible gases in chemical reactions. We had to see beyond the faculty of sight to *weigh* these invisible substances, and understand the causes behind these processes. He wrote:

> The true comprehension of air as a ponderable substance, and of gases in general as peculiar elastic and dispersive forms of matter, was only arrived at in the sixteenth and seventeenth centuries, and it was only after this that the transformations of substances could form a science. Up to that time, without understanding the invisible, but ponderable, gaseous and vaporous forms of matter, it was impossible to obtain any fundamental chemical knowledge, because the gases formed or used up in any reaction escaped notice.[3]

On the eve of the French Revolution, Antoine Lavoisier would unravel this mystery. Contemporary theory held that when burned, substances, including metals, lost a substance known as phlogiston, the "fire principle."

Changes in substances were explained by the addition or subtraction of phlogiston. In 1772, Lavoisier read the experiments of Guyton de Morveau, who showed that metals *increased* in weight when they were roasted in air. How could this be reconciled with the idea that burning was the removal of something? Although this did not bother the proponents of phlogiston theory, who explained it away by claiming that phlogiston can have "levity" which buoys up metals, it was a clear sign to Lavoisier that the theory was flawed. Lavoisier meticulously repeated the experiments, and found that when lead and tin were heated in closed containers their weights did not change; but when air was allowed to enter, the resulting product—the metal plus the burned ash—weighed *more* than the original metal.

He reasoned that some part of the air must be attaching itself to the metal. Soon thereafter, the chemist Joseph Priestley visited Lavoisier in Paris to tell him that he had found a new "dephlogisticated air" by heating up red calx of mercury (now called mercuric oxide, HgO). The new air seemed stronger and purer than regular air. Mice could live longer in the new air, than they would confined in an equal volume of regular air, and the new air

---
3. Mendeleev, Dmitri. *The Principles of Chemistry*, ed. A.J. Greenaway, trans. George Kamensky. London: Longmans, Green, and Co., 1891.

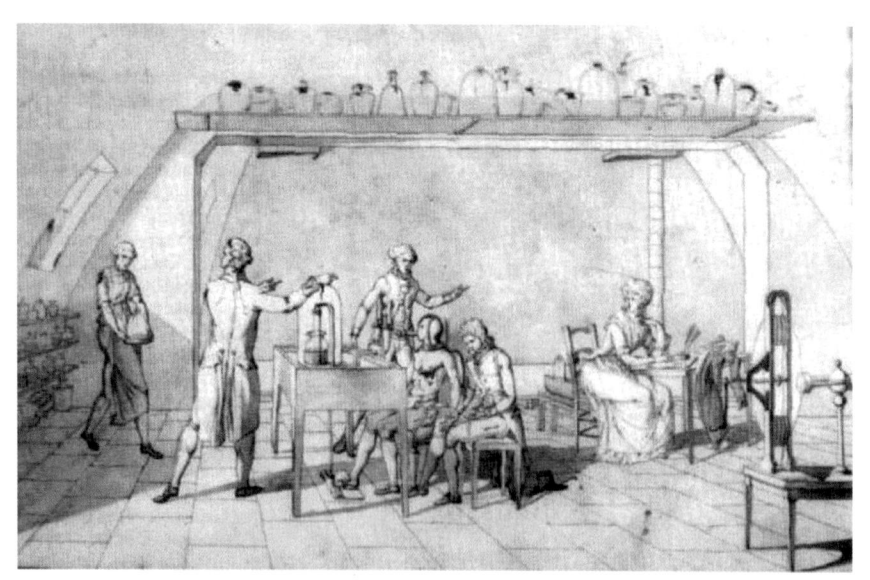

Monsieur and Madame Lavoisier and assistants experiment with respiration. Drawing by Madame Lavoisier, circa 1780.

allowed candles to burn with "an amazing strength of flame." Lavoisier repeated the experiment, and found the same result, but made a new hypothesis. Heating up the red calx of mercury had liberated something from it, and this substance was the same as that which was sticking to the heated lead and tin. Lavoisier identified this as an elementary substance, and later named it "oxygen."[4] He demonstrated that burning, rusting, and breathing were all types of oxidation—transformations in which oxygen combines with some other substance. Burning coal is rapid oxidation while rusting iron is slow oxidation.

Most gases burn, due to their ready combination with the oxygen in the air, in the presence of a flame. Hydrocarbons such as the methane in natural gas are eager to combine with oxygen, and burn quite well. After a tinsmith in Fredonia, New York in 1825 first observed bubbles forming in a creek, and decided to drill a well and sell the gas, the commercialization of natural gas as a fuel source took off.

So, what was the difference between these highly flammable natural gases, and the strange Dexter gas that refused to burn?

Return to Dexter

Using an air compressor and liquifier, the University of Kansas chemists were able to separate out the different gases. They found that it was only 15 percent methane, which was rendered non-flammable by 72 percent nitrogen. Along with the non-burning nitrogen was another 12 percent of a mysteriously "inert residue," out of which they were able to isolate, to their utter amazement—helium.

Helium wasn't supposed to be found in the Earth. At least not in the large quantities they had just discovered beneath the Great Plains. It was the Sun element, named from the Greek word for Sun—*helios*, where it was first observed, spectroscopically. Although it was quite a surprise to find helium on Earth, it was utterly useless as a fuel source since it did not burn, and for years, the entire U.S. supply of helium sat in three glass vials on a shelf at the University of Kansas.

Helium wouldn't burn, yet it was found in the Sun. Was the Sun not burning?

Helium was famous for being the first extraterrestrial element ever discovered. After the German physicist Gustav Kirchhoff figured out, in 1859, how to determine the chemical composition of stars by analyzing their light, astronomers eagerly anticipated the next total solar eclipse, so that they could analyze solar prominences. That opportunity came in 1868. French astronomer Pierre Jules César Janssen traveled to India with his spectroscope, and waited for the Moon to perfectly match the circumference of the Sun, blocking out the light of the bright orb, and leaving visible the protruding solar prominences.

Janssen observed a distinct yellow line in his spectroscope that was similar to the signature of sodium. Other scientists on the scene wrote it off as merely sodium, but Janssen thought it was a new element.

Meanwhile, in England, the English astronomer Joseph Norman Lockyer had figured out how to observe solar prominences in regular sunlight, and had also observed the bright yellow spectral line of the new element. Even though these two scientists, working independently,

---

4. Lavoisier named *oxygen* from Greek words meaning "acid maker." In the preface to his famous *Elements of Chemistry*, Lavoisier credits his advances in the science to his intention to improve chemical nomenclature:

"Thus, while I thought myself employed only in forming a nomenclature, and while I proposed to myself nothing more than to improve the chemical language, my work transformed itself by degrees, without my being able to prevent it, into a treatise upon the elements of chemistry. The impossibility of separating the nomenclature of a science from the science itself, is owing to this, that every branch of physical science must consist of three things: the series of facts which are the objects of the science; the ideas which represent these facts; and the words by which these ideas are expressed. Like three impressions of the same seal, the word ought to produce the idea, and the idea to be a picture of the fact. And, as ideas are preserved and communicated by means of words, it necessarily follows, that we cannot improve the language of any science, without at the same time improving the science itself; neither can we, on the other hand, improve a science, without improving the language or nomenclature which belongs to it. However certain the facts of any science may be, and however just the ideas we may have formed of these facts, we can only communicate false or imperfect impressions of these ideas to others, while we want words by which they may be properly expressed."

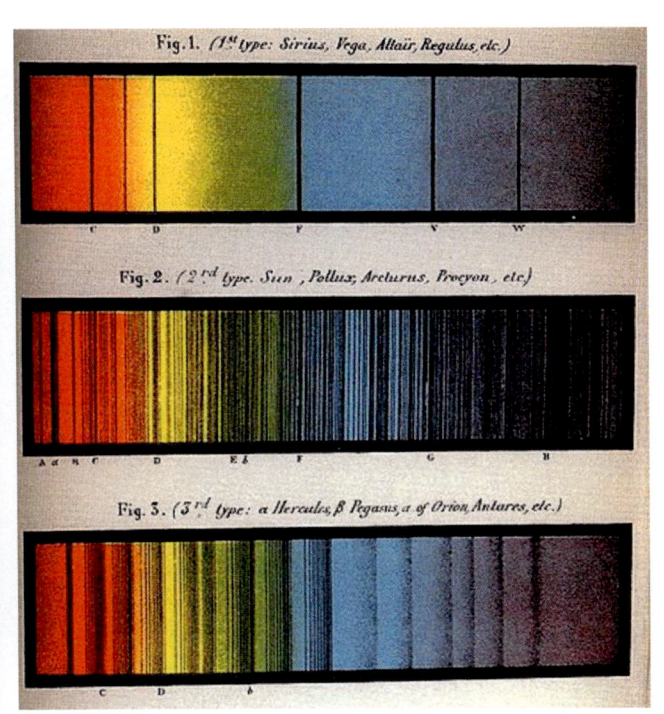

The spectroscope uses a prism to bend, or refract white light, which is made up of many different colors of light. Each color of light represents a unique wavelength and bends at a different angle, and the light spreads out, divided by color, into a broad rainbow. Joseph von Fraunhofer (1787-1826), a German telescope lens maker, used candle light to focus his lenses. A properly focused lens would not have a prismatic effect that spread the light out into distinct colors. One day, he used sunlight to focus his lenses, instead of a candle, and noticed some strange black lines in the spectrum. He figured out that the different lines represented different elements that were in the Sun. The black lines indicated certain wavelengths of light that were being absorbed by certain elements. Each element would absorb a series of wavelengths, which formed a pattern–a characteristic signature for each element. Depending on how this spectrum is observed, either a continuous spectrum of light can be seen, with breaks of black lines, where certain frequencies are absorbed, or the inverse–only lines of color, where those same frequencies are emitted.

5,000 miles apart, had come to the same conclusion, and were able to register their discoveries on the very same day at the French Academy of Sciences, they received little acclaim. The spectral results could not be reproduced in a lab, and no one believed that this new alien element existed.

They would not receive due credit until almost 30 years later, when helium would, again, emerge in a very mysterious process.

Alpha Particles

Marie and Pierre Curie spent endless hours investigating the strange properties of certain minerals that emitted a new form of energy. Henri Becquerel had previously found that uranium salts radiate a type of invisible light that can expose photographic plates. Marie Curie experimented with different compounds of uranium and thorium and noticed that no matter what type of minerals these special elements were found in, they all emitted the radiation in the same way.

This did not fit the proper behavior of chemistry. Compounds of the same element often possess very different chemical properties. For example, one compound of uranium can be a dull black powder, while another can be a clear yellow crystal that glows green. Marie Curie found that the only thing that affected the amount of radiation emitted was the amount of uranium or thorium that the compound contained. She thus reasoned that this radiation was not the result of a *chemical* property, i.e., an effect of the different atoms' structural arrangement and relationship between each other. She hypothesized that radiation must originate from *inside the atom itself*.

After discovering radium, which was one million times more radioactive than uranium, the Curies put radioactivity to the test, poking and prodding these elements to figure out the nature of this new energy. The influence of a magnetic field revealed that the radiation was composed of different types of rays, some of which were affected by magnetism. When physicist Ernest Rutherford repeated the experiment, using an even stronger magnetic field, he was able to find three distinct rays.

The first type of rays were clearly and narrowly bent. The second type were more strongly bent and spread out

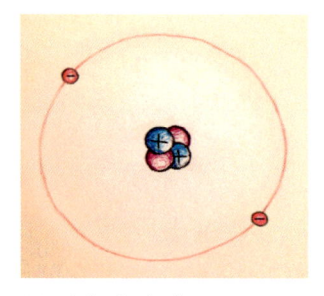

Model of a helium-4 atom. 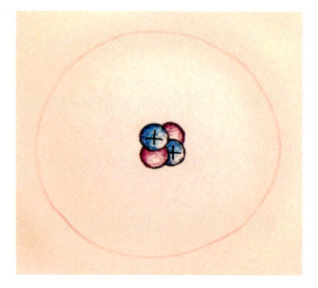 *Doubly ionized helium-4.*

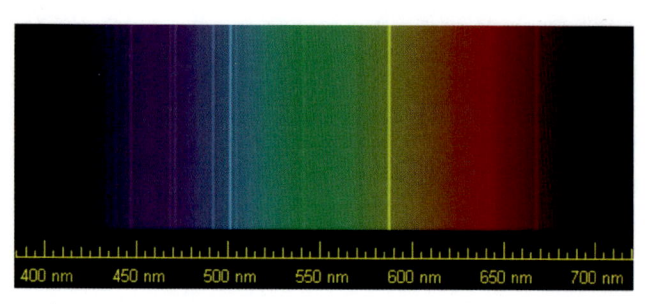

Helium emission spectrum.

in a broad band. The third type was not affected at all by the magnet, and kept on its straight and narrow course. These rays were called alpha, beta, and gamma rays, respectively. Alpha were electrically positive, beta negative, and gamma neutral.[5] Rutherford found that the beta rays were electrons, and the alpha rays were a stream of oppositely charged, much heavier particles. Based on their weight and charge, he hypothesized that alpha particles were doubly ionized helium atoms—i.e., they were helium atoms which had lost both of their electrons, leaving no electrons, and only a bare, positively charged nucleus.

This hypothesis was corroborated by the then-recent discovery that most radioactive mineral ores contained helium atoms.

In 1895, the Scottish chemist William Ramsay heard that a Norwegian mineral called cleveite[6] emitted a gas similar to nitrogen when it was heated. Having discovered argon the year before, which had also been mistaken for nitrogen by other scientists,[7] Ramsay decided to treat the cleveite with sulfuric acid, to find out if argon would be liberated from it. When he examined the gas, Ramsay was so surprised by the bright yellow line that appeared on his spectroscope, that he thought he must be misreading it, and proceeded to clean his instrument. He then sent the gaseous emanation to Lockyer to identify. It was not argon, but a new terrestrial element, which matched the same yellow signature of Janssen and Lockyer's alleged Sun element, helium.

---

5. See "The Nuclear Era: Man Controls the Atom" in this report.

6. Cleveite is a radioactive variety of uraninite, with composition UO_2, where about 10% of the uranium is replaced by rare earth elements.

7. In 1892, Lord Rayleigh could not make sense of the very slight discrepancies in his measurements of nitrogen in the air, and wrote a plea to other scientists in *Nature*: "I am much puzzled by some recent results as to the density of nitrogen, and shall be obliged if any of your chemical readers can offer suggestions as to the cause." Mendeleev's periodic table had been established in 1869, and there were no empty spaces for an element of this type. William Ramsay made the bold hypothesis that there might be a whole new *family* of elements, and that the discrepancy was due to a heavier element of this new family, hidden in the air. He was correct. His discovery of argon was the first element of a new column of inert elements—the noble gases. Lord Rayleigh, "Density of Nitrogen," *Nature* 46, 512 (1892).

How strange that a substance that did not form molecules could be found inside so many minerals.[8] How did it get inside these minerals, if it does not like to bond with anything? Why was this chemically useless element found in radioactive minerals? Was the Sun somehow implanting radiation in rocks?

William Ramsay and Frederick Soddy observed the radioactive gases with a spectroscope over time to see if they could figure out the nature of the transformations occurring.

They collected gaseous emanations from radium, and sealed them in a tube, through which a current was passed. The gas emitted light, whose spectrum they could observe, and to their surprise, over time, the spectral lines changed. The lines of radium emanation glowed with less intensity, and as they faded, a new bright yellow one emerged. The radium emanation was actually being transformed into another element. Helium was being created from radium. This confirmed Marie Curie's hypothesis, that this was not a chemical process, but a change occurring, inside the atom, on a nuclear level—i.e., the generation of new elements came out of the transformation of the atomic nucleus.

Helium would not partake in chemical reactions because it had a different identity—an identity as a future artifact of the nuclear era, and beyond.

Beyond Chemistry

This odorless, colorless, tasteless, chemically worthless lighter-than-air element was useless before the advent of modern science. But as we made the societal advances that allowed for the development of the native resources of the mind, the inherent qualities of this element would begin to manifest themselves.

The belief that helium was an extraterrestrial element was more prescient than those nineteenth-century astronomers—who named it after the Sun—understood at the time.

---

8. Helium does not form molecules, burn, or chemically react with other elements because it does not share outer electrons with other atoms. The sharing of outer electrons is what constitutes chemical change. Helium only has two valence electrons, which is considered a full, stable shell, and it is not inclined to share.

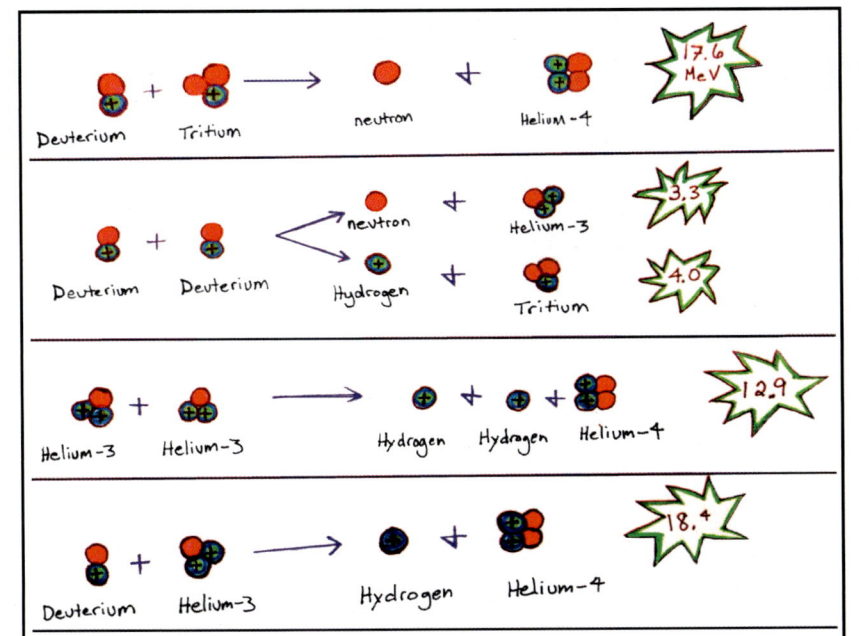

Four different fusion reactions, involving deuterium (D), tritium (T) and helium-3 as fuels. Output products are shown, along with energy released per reaction, expressed in MeV. The D-D reaction has two possible outputs. Neutrons cannot be affected by a magnetic field, although the other (charged) particles can. Helium-3 fuel makes it possible to have reactions without neutrons. Note that combining D and He-3 fuel will also result in D-D fusions, and will therefore produce some neutrons.

This second most abundant element in the universe, which escaped our grasp until almost the twentieth century, also almost escaped from the planet, until legislation was introduced in 1958 to capture and conserve it. The helium that is created from the radioactive decay of heavy elements deep in the earth's crust makes its way out of the ground, and being lighter than air, has nothing to keep it in the atmosphere, so it escapes into space. That recognition, scientifically and politically, would allow helium to take us off the Earth,[9] and all the way to the Moon. Its very low freezing point would make it the only thing that could be used as a refrigerant for liquid oxygen and hydrogen rocket fuels. During the Apollo program, helium would determine how long the astronauts could stay on the Moon. Once the helium had boiled away, there would have been nothing left to keep the return fuel in liquid form, and the spacecraft would have been stranded.

It would continue to prove its worth in advanced technologies due to its ability to be cooled almost to absolute zero while still remaining a liquid,[10] and is therefore used for superconducting magnet technology, magnetic resonance imaging, and advanced cryogenic research.[11]

But helium has an even nobler mission in advanced sciences—and the future of human civilization—that is yet to be met. The even more extraterrestrial identity of helium's special isotope, helium-3 will be vital for helping us achieve our own extraterrestrial imperative.[12]

Helium-3 Fusion: A New Type of Energy

The hidden potential of this ethereal isotope currently resides in a domain beyond the chemical, beyond nuclear fission and beyond even many nuclear fusion reactions. Fusion reactions involving helium-3 are considered advanced, third generation reactions due to the relative difficulty in achieving them with current magnetic confinement technologies. Helium-3 fusion reactions are truly advanced due to the qualitative power increase that they represent, compared to all other current forms of energy production.

Since the modern era of electricity production began with the advent of the steam powered turbine in 1884, the primary source of energy has been based on rotary motion to drive an electrical generator. Today, approximately 90% of all electricity generation in the United States is by use of a steam turbine. Each successive stage of higher energy-flux density fuel sources—coal, natural gas, nuclear fission, and nuclear fusion—represent advances in the potential of that fuel, as measured in the relative quantity of the material to its energy output. Although the density of energy innate to each of these fuel sources is different, the *type* of energy generated remains the same: heat. In each of these processes, we are merely using a

---

9. Its lighter-than-air, non-flammable properties would make it a key resource to the U.S. Navy during WWII for its use in surveillance blimps to detect German submarines. The Germans' lack of helium forced them to use highly flammable hydrogen in the unfortunate *Hindenburg*.

10. Helium boils at 4.22 Kelvin or −452 degrees Fahrenheit.

11. Helium-4 also has a very strange "quantum state," that defies the laws of classical physics. Once it reaches a special liquid state at 4.2 K, it gains properties such as zero viscosity, which allows it to literally crawl up walls, and imitate the properties of sound. What new principles lie dormant, awaiting us to uncover them? What future potential does this hint at? See Alfred Leitner's 1963 video demonstrating these properties at alfredleitner.com

12. German rocket propulsion engineer and space pioneer Krafft Ehricke (1917–1984) believed that human creativity possessed no limits, and that as a uniquely creative species we have an "extraterrestrial imperative" to explore and develop space in order for the species—and that creative quality—to progress.

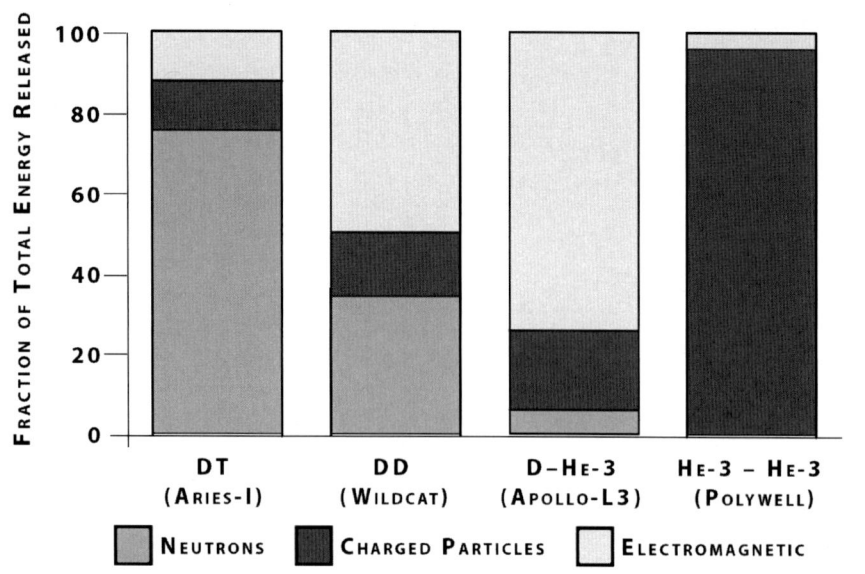

Fusion reactions release energy, and that energy can come in three forms: the motion of neutrons, the motion of charged particles, and in electromagnetic radiation (forms of light). This diagram indicates energy release breakdowns for several proposed fusion designs.

different fuel source to generate energy of motion (kinetic energy), which heats up water to create steam, to spin a turbine in a magnetic field, to induce an electric current.

Helium-3 fusion reactors offer the potential to liberate us from this 130-year old technology, and move us into the next era.

When the nuclei of light atoms are forced together in the process of controlled thermonuclear fusion, they make different products. Among those products can be positively charged particles, neutral particles, and different types of electromagnetic radiation. These charged particles, neutrons, and photons serve different purposes for energy production.

A *first generation* fusion reaction involves two isotopes of hydrogen—deuterium and tritium (DT). When these isotopes fuse, the reaction creates 80% neutrons, along with photons and some charged helium nuclei (alpha particles). The energy from this reaction is taken from the kinetic motion of the high-energy neutrons. Although the energy density of this fusion fuel is higher than in fission reactions, the same physical process is at play. High energy neutrons create heat, which must be converted into electricity. Furthermore, because neutrons are neutral, i.e., they have no charge, they do not respond to a magnetic field, and are thus very hard to control. These factors, combined, give the DT reaction an electrical conversion efficiency of 45%, not much better than a fission reaction (40%), or any heat-based form of electrical energy for that matter.

A *second generation* reaction, using helium-3 and deuterium, generates very different fusion products. In this case, depending on factors such as plasma temperature and the ratio of helium-3 to deuterium, hardly any neutrons (1-5%) will be produced, and the majority of the products will be in the form of charged particles (protons and alpha particles) and photons. Instead of having to convert the heat generated from neutrons into electricity, the charged particles and electromagnetic radiation are directly converted to electricity. Direct conversion methods yield efficiencies of 60-70%.

The main advantages of these products, as opposed to neutrons,[13] is the greater ease in directly converting them to electricity, and the fact that charged particles *do* respond to a magnetic field, and can thus be efficiently controlled and directed.[14]

Magnetohydrodynamics is one method for using this flow of charged particles to generate electricity directly. A moving charge under the influence of a magnetic field, will be deflected. By passing a charged particle plasma (which conducts current) through a magnetic field, the charge is deflected to one side by the magnetic field, creating a potential difference and the flow of current.

Electrostatic direct conversion makes electricity by creating voltage—the electrical potential difference between two points—from the motion of the charged par-

---

13. It should be noted that neutrons are not inherently bad things. They can be very useful for certain purposes, such as the production of life-saving medical isotopes, or for explosive detection technologies. In a process such as desalination, where heat may be used for evaporation, we may prefer a neutron-producing fusion or fission process that can both generate electricity, while using waste heat for the desalination process.

14. While first generation DT reactions are thus classified because they are considered the easiest to achieve in terms of the temperature, pressure and confinement times required for magnetic confinement fusion, this practical approach (often a response to budget cuts and bad economic policy) may not be the fastest way to achieve commercial fusion, after all. A side effect of using an aneutronic helium-3 reaction is that we will avoid the extra engineering, maintenance and fuel-processing challenges that come with the nuclear radiation of DT reactions. We will not have to deal with the high-energy, out-of-control neutrons that wreak havoc on reactor walls and other metallic components, and require radiation shielding and cooling towers. By eliminating the time and expenses required to develop these materials, we may concentrate our resources on plasma physics.

ticles. While a particle accelerator uses voltage differences to induce motion in particles, this process works in reverse, using the motion of the charged particles created by the fusion reaction to drive the voltage. In effect, the charged particle is slowed electrostatically, during which process it drives a current.

An advantage of electromagnetic products is that this radiative energy can be tuned to make use of specific wavelengths. Microwaves, gamma rays and X-rays may be selected and used for various applications aside from electricity. There are also methods for converting radiative energy into electricity.

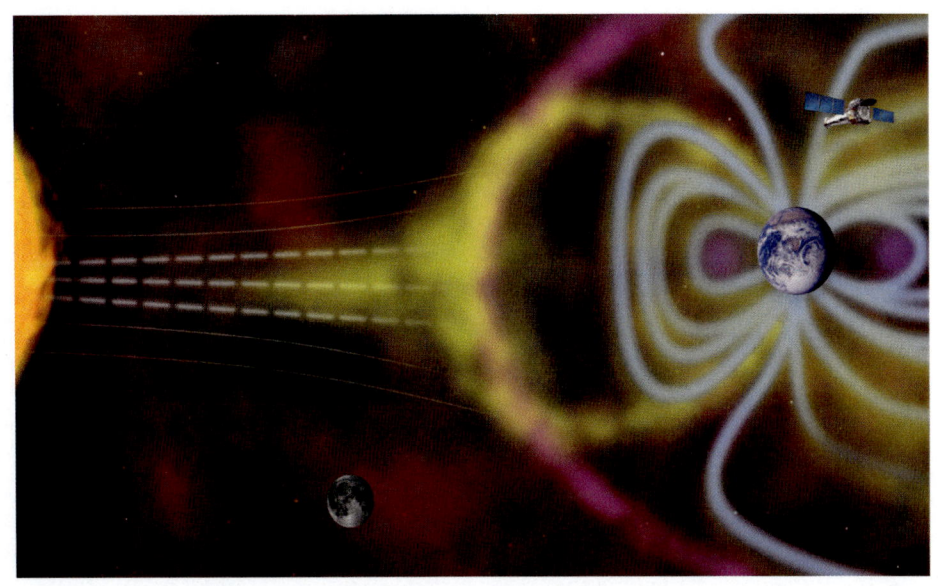

Artist's vision of the Earth's magnetic field, protecting our planet from the charged particles in solar wind, while the exposed Moon is subject to the full brunt of solar emissions, including the beneficial fuel helium-3.

One method uses a rectifying antenna called a "rectenna" to convert microwave energy into direct current electricity. The inventor of this device, William C. Brown reported to NASA's Second Beamed Space-Power Workshop in 1989 that he had demonstrated an 85% electricity conversion efficiency.[15]

A *third generation* fusion reaction uses helium-3 as both agents in the reaction. In an electrostatic device,[16] 99% of the resulting energy is in charged particles, which can be directly converted into electricity, yielding an electrical conversion efficiency of 70-80%. There are no neutrons or radioactivity produced in a He-3–He-3 reaction.[17]

Finding Helium-3

When fusion scientists at the University of Wisconsin's Fusion Technology Institute realized the value of helium-3 for nuclear fusion reactions, they wondered where it could be obtained. Unlike the regular helium-4, which was discovered to be common by the Kansas chemists, helium-3 was still believed to be quite rare—at least on Earth. Then, they remembered that the Sun, a giant nuclear fusion reactor, was pumping out quite a bit of helium-3, as a product of fusing hydrogen. The Sun spews out helium-3 along with other charged particles and plasma into the solar system, in the form of solar wind and coronal mass ejections. On Earth, we're largely shielded by an atmosphere and a strong magnetic field. But our less fortunate Moon is completely exposed to all of the Sun's tantrums. The Wisconsin fusion scientists made the hypothesis that helium-3 could be found on the Moon. In 1986, they made a trip down to NASA's Johnson Space Center in Houston, to scour the records of Apollo lunar samples.

Indeed, records showed helium-3 to be present in every lunar sample.

Lunar scientists whom they queried about the rare isotope were puzzled. They said that they had known since 1970 that there was an abundance of helium-3 on the Moon, but were not aware that it was useful for anything. Of course, it was not useful for anything in 1970, because the discovery of its vital importance as a fusion fuel had not yet been made. The helium-3 lunar samples had been destined to sit, useless, on shelves at NASA, as had the Dexter gas at the University of Kansas. And it will remain seated on the lunar shelves of our natural satellite, the Moon, until there is a significant breakthrough made here on Earth.

A serious step in that direction has been made by the Chinese with their December 14, 2013 landing of a spacecraft on the Moon. While we do not have full access to the plans of the Chinese, we do know something about their intentions, and the technical capabilities that have been made possible by the pioneering work of scientists at the Fusion Technology Institute of the University of Wisconsin, and the Department of Earth and Planetary Sciences, at the University of Tennessee, since the U.S last visited the Moon, in December 1972. Research-

---

15. Freeman, Marsha, "Mining Helium on the Moon to Power the Earth" *21st Century Science & Technology*, Summer 1990.

16. Kulcinski, G.L. and Schmitt, H.H., "Nuclear Power Without Radioactive Waste—The Promise of Lunar Helium-3," 2000.

17. Kulcinski, G.L. "Helium-3 Fusion Reactors—A Clean and Safe Source of Energy in the 21st Century," 1993.

ers found that the helium-3 is held very loosely in the dust on the surface of the moon and could be extracted relatively easily. Scientists at the Wisconsin Center for Space Automation and Robotics have designed vehicles to separate helium-3 from the lunar soil. If it is heated to 600-700°C, it can be released from the dust and re-cooled into a liquid during the cold lunar night. This can be done by concentrating solar energy with mirrors, or by using microwave energy, which has a very unique coupling effect with lunar soil, that allows it to be heated very efficiently with microwave energy. The potential reserves of helium-3 are estimated at one million tons, which could power the Earth in fusion reactors for 1,000 years. It also has been shown that there is ten times more energy in He-3 on the Moon than there ever was in fossil fuels (i.e., coal, oil, and gas) on the Earth. This fossil of the Sun is magnitudes more energy dense than any petroleum product, such that one shuttle load could supply the entire U.S. with electricity for one year.[18]

Chinese Moon goddess, Chang'e. Chinese President Xi Jinping, in a speech to space scientists and engineers who participated in the research and development of the Chang'e-3 mission, said that innovation in science and technology must be put in a "core position" in the country's overall development: "Dare to walk the unwalked paths. Constantly seek excellence through solving difficulties, and accelerate the shift to innovation-fueled development."

---

18. This was measured in 1988, when the U.S. still operated the Space Shuttle. Our electricity consumption is not much higher than 1988 terms, due to economic collapse, and the resulting reduction of industry.

The development of helium-3 fusion reactors on the Moon would give us a unique power for industrial and agricultural applications that could take advantage of the low gravity, near vacuum, extreme temperature changes, and other conditions. This is an ideal fuel for use on the Moon and other space applications, because it is available on site, and because the direct conversion to electricity mitigates any thermal losses.

For every ton of helium-3 extracted, there are 6,000 tons of hydrogen, 500 tons of nitrogen, 5,000 tons of carbon-containing molecules, and over 3,000 tons of the heavier helium-4 isotope, all of which will be extremely valuable for atmospheric control, life support, and chemical fuels during the construction of a lunar base.

Fusion rockets far exceed the energy-flux density of chemical rockets, allowing for much less fuel mass, and, crucially, making it possible to fly missions that simply could not be undertaken with chemical propulsion, such as one-week transit time to Mars (instead of many months), and an effective strategy for planetary defense.[19]

Among fusion fuels, helium-3 is by far the best, because the products of helium-3 fusion reactions are mostly charged particles, creating a magnetically controlled exhaust to propel the rocket. As stated by fusion scientist John Santarius, "Fusion will be to space propulsion what fission is to the submarine."

While the isotope helium-3 is much more rare on Earth than helium-4, we do have access to a small amount that could be used to build test facilities. Although using the natural helium-3 left over from the formation of the Earth would require extracting all natural gas in the planet, and would only yield 200 kg, there is another source. Both the United States and Russia have about 300 kg worth that could be collected from the radioactive decay of tritium in thermonuclear weapons. This would be more than enough to fuel test facilities to develop the proper fusion engineering to get us started.

How to Find Helium-3 on the Moon: A New Spectroscopy

In order to begin a proper mining expedition, we will need to create a map of the Moon, which shows the locations of the higher concentrations of helium-3. Unlike on Earth, where there are veins of ores which have been concentrated by efficiently active forms of life, the resources on the Moon are more diffuse. However, since it is the Sun that is implanting the helium-3, we can know that there will be more helium-3 in the places where the Sun has been able to reach more easily, i.e., the surface. This is a very fortunate situation, since it means we will

---

19. See the Planetary Defense issue of *21st Century Science & Technology*, Fall/Winter 2012–2013.

not have to embark on complex drilling missions deep below the surface of the Moon.[20] Because the Sun does not affect the Moon's surface uniformly, the distribution of helium-3 is also non-uniform. We can use this non-uniform behavior of the Sun to detect where there will be greater amounts of helium-3.

We can do this using gamma ray spectroscopy to detect when the Sun creates changes in the helium-3 that is embedded in the surface of the Moon. Researchers at the Fusion Technology Institute, propose to use very large solar proton flares, to take advantage of the increased flux of solar cosmic-ray–induced neutrons.[21] When neutrons from these solar flares reach the surface of the Moon, they can react with helium-3, and that reaction can be detected.

The difference between helium-3 and helium-4 is that fourth thing, the extra neutron. When helium-3 is bombarded with a neutron and is transformed into helium-4, a little burst of energy is produced, in the form of a gamma ray.[22]

Gamma rays also have signatures like the distinctly colored spectral lines characteristic of elements that can be seen with a spectroscope. These signatures depend on the amount of energy that the gamma ray has. The gamma ray that is produced from a reaction between a helium-3 atom and a neutron is a very specific energy—20.6 MeV—which is such a different value than that produced in other reactions, that it is not easily confused. While these reactions are infrequent, the specificity of that particular 20.6 MeV gamma ray can be uniquely detected. "We are essentially 'looking for a needle in a haystack.' Fortunately, it is a different colored needle."[23]

A gamma ray spectroscope can thus be used in a satellite orbiting the Moon, which will wait for these solar flares to instigate gamma-ray-releasing reactions with the helium-3. This is only one proposal for creating a map of the Moon to mine this necessary new resource. With the international Apollo crash program to develop fusion energy that must be implemented before this decade is out, there will be many more.

---

20. The mining of this new resource, helium-3—magnitudes more energy-dense than petroleum—will be far easier in this respect than oil beneath the ocean floors, which must use NASA space technology to carry out increasingly complicated missions.

21. Karris, K.R., H.Y. Khater, G.L. Kulcinski "Remote Sensing of Astrofuel" 1993, Wisconsin Center for Space Automation and Robotics.

22. Remember, these were the third type of rays (alpha, beta, gamma) observed by Rutherford and the Curies, that constituted radioactive emanations. Gamma rays were the very high energy, fast, penetrating rays that were not swayed by the magnetic field.

23. Karris, K.R., H.Y. Khater, G.L. Kulcinski "Remote Sensing of Astrofuel" 1993, Wisconsin Center for Space Automation and Robotics.

Conclusion

It is estimated that as a result of fusion processes for the past four billion years, the Sun is now composed of about one-third helium, and has only two-thirds left of its original hydrogen.

As the Sun converts that remaining two-thirds hydrogen into helium and implants it into the Moon for storage, it is gradually losing its ability to create fusion reactions, and therefore losing its power as our Sun. The remainder of our Sun's life is estimated at approximately two billion years, which should give us enough time to recreate its processes. Retrieving from the Moon these helium fossils of the Sun's short life, and employing them to venture out into a new planetary system, so that we may survive to extend our creative reach into new worlds, is not a mission that can be delayed.

We must ask again, what is the value of helium, or any resource? Do resources exist independently of the human mind, and of a culture and economy that has chosen to discover and make use of them? Is economic value really a function of money? Would all the money gained through the imperial wars of Zeus, from the Temple of Delphi to the present day, have been sufficient to build a helium-3 fusion reactor in those times, under those systems? Does our species have the collective moral intelligence at this moment to cast off the Zeusian shackles of our slow development and soar, before it is too late?

"China has made no secret of their interest in lunar Helium-3 fusion resources." Former astronaut, geologist and U.S. Senator, Harrison Schmitt is one of the leading proponents for the mining of helium-3 on the Moon. He was on the last Apollo mission to the Moon.

Appendix

Prometheus
The Historical Record

by Jason Ross

Prometheus was a historical personality, who endured the wrath of the Zeus for daring to bring "fire" and science to man. The oldest direct historical knowledge of Prometheus comes from the Greek poet Hesiod in his *Theogony* and *Works and Days* and the Greek playwright Aeschylus in his play *Prometheus Bound*.[1] The story they relate of Prometheus the Fire-Bringer is one that finds parallels in other cultures, and may actually date back to the early part of the Bronze Age.[2] His story served as inspiration for works by Percy Bysshe Shelley, who wrote *Prometheus Unbound* and Johann Wolfgang von Goethe, who wrote a poem *Prometheus*, and had intended to compose an entire play.

According to Hesiod and Aeschylus, Prometheus was one of the Titans, the ruling group of immortals that pre-dated the gods of Olympus. Zeus, with the help of Prometheus, overthrew Kronos, the ruler of the Titans, to become the ruler of the gods of Olympus.

Before the action related in Aeschylus's play, Prometheus had acted to benefit mankind. First, he had established a tradition of sacrifices by a trick he played on Zeus. Cutting apart an ox, Prometheus separated out two piles: one of meat and organs, wrapped in the ox's unsightly stomach, and another of the bones, carefully covered with shiny fat. Prometheus asked Zeus to choose which pile he would accept as a sacrifice. Zeus chose the fat-covered bones, starting the customary sacrifice of bones and fat to the gods, while keeping the meat for mankind. Zeus, enraged, refused fire to man as punishment. Prometheus saved man from this fate. He stole fire from heaven, and gave it to man. For this, he received the full wrath of Zeus (and the anger of many of the other gods).

As Aeschylus's play opens, Prometheus is being conducted to a desolate rocky crag, where he is to be bound as long as Zeus's anger lasts. With a stake driven through his chest, pinning him to the rock, and his arms and legs bound, Prometheus is to suffer the endless torment of having his liver devoured every day by an eagle (a symbol of Zeus), only to have it grow back each night.

After he is bound, a Chorus of the daughters of Oceanus flies to his location, to speak to him. Aeschylus writes:

CHORUS: Unfold the whole story and tell us upon what charge Zeus has caught you and painfully punishes you with such dishonor. Instruct us, unless, indeed, there is some harm in telling.

PROMETHEUS: It is painful to me to tell the tale, painful to keep it silent. My case is unfortunate every way....

You ask why he torments me, and this I will now make clear. As soon as he had seated himself upon his father's throne, he immediately assigned to the deities their several privileges and apportioned to them their proper powers. But of wretched mortals he took no notice, desiring to bring the whole race to an end and create a new one in its place. Against this purpose none dared make stand except me—I only had the courage; I saved mortals so that they did not descend, blasted utterly, to the house of Hades. This is why I am bent by such grievous tortures, painful to suffer, piteous to behold. I who gave mortals first place in my pity, I am deemed unworthy to win this pity for myself, but am in this way mercilessly disciplined, a spectacle that shames the glory of Zeus.

CHORUS: Iron-hearted and made of stone, Prometheus, is he who feels no compassion at your miseries. For myself, I would not have desired to see them; and now that I see them, I am pained in my heart.

PROMETHEUS: Yes, to my friends indeed I am a spectacle of pity.

CHORUS: Did you perhaps transgress even somewhat beyond this offence?

PROMETHEUS: Yes, I caused mortals to cease foreseeing their doom.

---

1. Hesiod was active around 700 BC, and Aeschylus fluorished in the fifth century BC.

2. Sulek, Marty, "Mythographic and Linguistic Evidence for Religious Giving among Graeco-Aryans during the Chalcolithic Age," presented at the July 2012 ISTR conference in Siena, Italy.

CHORUS: Of what sort was the cure that you found for this affliction?

PROMETHEUS: I caused unseen hopes to dwell within their breasts.

CHORUS: A great benefit was this you gave to mortals.

PROMETHEUS: In addition, I gave them fire.

CHORUS: What! Do creatures of a day now have flame-eyed fire?

PROMETHEUS: Yes, and from it they shall learn many arts.

Further on, Prometheus continues:

PROMETHEUS: Still, listen to the miseries that beset mankind—how they were witless before and I made them have sense and endowed them with reason. I will not speak to upbraid mankind but to set forth the friendly purpose that inspired my blessing.

First of all, though they had eyes to see, they saw to no avail; they had ears, but they did not understand; but, just as shapes in dreams, throughout their length of days, without purpose they wrought all things in confusion. They had neither knowledge of houses built of bricks and turned to face the sun nor yet of work in wood; but dwelt beneath the ground like swarming ants, in sunless caves. They had no sign either of winter or of flowery spring or of fruitful summer, on which they could depend but managed everything without judgment, until I taught them to discern the risings of the stars and their settings, which are difficult to distinguish.

Yes, and numbers, too, chiefest of sciences, I invented for them, and the combining of letters, creative mother of the Muses' arts, with which to hold all things in memory. I, too, first brought brute beasts beneath the yoke to be subject to the collar and the pack-saddle, so that they might bear in men's stead their heaviest burdens; and to the chariot I harnessed horses and made them obedient to the rein.... It was I and no one else who invented the mariner's flaxen-winged car that roams the sea. Wretched that I am—such are the arts I devised for mankind, yet have myself no cunning means to rid me of my present suffering.

CHORUS: You have suffered sorrow and humiliation. You have lost your wits and have gone astray; and, like an unskilled doctor, fallen ill, you lose heart and cannot discover by which remedies to cure your own disease.

PROMETHEUS: Hear the rest and you shall wonder the more at the arts and resources I devised. This first and foremost: if ever man fell ill, there was no defense—no healing food, no ointment, nor any drink—but for lack of medicine they wasted away, until I showed them how to mix soothing remedies with which they now ward off all their disorders... Now as to the benefits to men that lay concealed beneath the earth—bronze, iron, silver, and gold—who would claim to have discovered them before me? No one, I know full well, unless he likes to babble idly. Hear the sum of the whole matter in the compass of one brief word—every art possessed by man comes from Prometheus.

Aeschylus (fifth century BC), the Greek playwright who wrote Prometheus Bound.

These gifts of Prometheus have been the subject of this report. Aeschylus's play continues with Zeus sending Hermes to demand that Prometheus repent for his actions and share his secret:

HERMES: Bend your will, perverse fool, oh bend your will at last to wisdom in face of your present sufferings.

PROMETHEUS: In vain you trouble me, as though it were a wave you try to persuade. Never think that, through terror at the will of Zeus, I shall become womanish and, with hands upturned, aping woman's ways, shall importune my greatly hated enemy to release me from these bonds. I am far, far from that.

Prometheus refuses, and in Zeus's rage, is swallowed in lightning, earthquake, tempest, and storm. So ends Aeschylus's first play of the Prometheus trilogy.

The other two plays, *Prometheus Unbound*, and *Prometheus the Fire-Bringer*, are lost, yet some aspects of the plot to come are known. *Prometheus Unbound* includes Heracles killing the eagle that has fed on Prometheus's liver, and freeing him from his chains. Zeus frees the other Titans he has imprisoned. And in *Prometheus the Fire-Bringer*, Prometheus reconciles with Zeus, informing him of what was to have been his downfall.

Modern Prometheus

While the title of "modern Prometheus" is applied to Benjamin Franklin, whose work on electricity garnered him almost as much early renown as his later work on American independence, several notable modern treatments of Prometheus differ from that of Aeschylus.

Goethe's Prometheus shows nothing but contempt and scorn for Zeus. It is man's own actions that bring him advancement, not plaintive wishes to the uncaring heavens. "I know nothing shabbier under the sun than ye gods!" he exclaims, asking "I revere thee? What for?"

Goethe's Prometheus concludes: "Here I sit, forming humans / In my own image, / It will be a race like me, / For suffering, weeping, / Enjoying and rejoicing, and shall / Pay thee no attention, / Like me!"

Percy Bysshe Shelley similarly allows for no reconciliation between the worst of tyrants and the greatest of benefactors. He writes in the preface to his play *Prometheus Unbound*:

> The *Prometheus Bound* of Aeschylus supposed the reconciliation of Jupiter [Zeus] with his victim.... Had I framed my story on this model, I should have done no more than have attempted to restore the lost drama of Aeschylus.... But, in truth, I was averse from a catastrophe so feeble as that of reconciling the Champion with the Oppressor of mankind. The moral interest of the fable, which is so powerfully sustained by the sufferings and endurance of Prometheus, would be annihilated if we could conceive of him as unsaying his high language and quailing before his successful and perfidious adversary.

There can be no compromise with tyranny. Shelley presents us, in Act I, with Prometheus's captivity, his curse upon Zeus, the attempts of Hermes to persuade him to reconcile himself with Zeus's power, and Zeus's anger when he absolutely refuses. Shelley concludes the play with an epilogue addressing Prometheus (the "Titan" referenced below):

> To suffer woes which Hope thinks infinite;
> To forgive wrongs darker than death or night;
> To defy Power, which seems omnipotent;
> To love, and bear; to hope till Hope creates
> From its own wreck the thing it contemplates;
> Neither to change, nor falter, nor repent;
> This, like thy glory, Titan, is to be
> Good, great and joyous, beautiful and free;
> This is alone Life, Joy, Empire, and Victory.

We must not falter as we expand the gifts of Prometheus, and eliminate the oligarchical forces that have prevented man from being fully Promethean. As we consider the stunning developments made in physical chemistry, we must not forget that these advancedments have been hated and opposed by Zeusians along the entire history of the human species, most recently through the Anglo-Dutch empire, the depopulation intention of Queen Elizabeth and her cohorts, and the plague of anti-humanism masquerading as concern for the environment.

Today's Zeusians would rather have global thermonuclear war than cede their control to Promethean economic development. The reign of Zeus must come to an end.

Lyndon LaRouche on Prometheus

In his "Mind Over Your Matter," Lyndon LaRouche contrasts the Zeusian and Promethean outlooks:[1]

> Zeus forbad "fire's use" by mankind; Prometheus demanded the use of fire by mankind (which is the distinction of man from ape)... What Prometheus intended, and there was only one particular error-of-omission in this matter on his part: is what modern science knows under the caption of modern "chemistry:" which, in turn, is, in-fact-of-practice, best exemplified by mankind's successive increase in the energy-flux-density of the upward course of the evolution of chemistry: as that lies under the essential inclusion of the leading factor of human progress centered in the use of "fire:" increasing leaps in the quality of "fire," per-capita, and per-leap in the application of increasingly concentrated energy-flux-densities.

Later, he quotes himself from his February 7, 2014 webcast event, giving a beautiful, guiding concept of being Promethean:

> Actually, creativity is located within the ability of the individual to make a discovery of a principle of nature, and that's chemistry. Mankind operates on the basis of chemistry! It's called fire! It's the Promethean force of fire! [I.e., energy-flux density.] And, by fire, you rise to higher and higher powers of chemistry. And, many chemists get confused on this, because they get so trapped up with [what] they can do, they forget about discovery [of principles]. But, everything that was done in chemistry came about as a discovery! A creation of the human mind, of the human imagination! And, the ability to criticize your imagination, the human imagination! And, the ability to criticize your own imagination, and to determine, by testing, whether this thing you call a principle, is true, or not: you test it.
>
> Now, many chemists don't do it properly; but, the intention in the system of chemistry, is there. Mankind is the fire-bringer! He's the Promethean, the fire-bringer! And, what he's doing is discovering new, higher forms of fire: like the application of helium-3 to the process of creating a superpower for mankind, per capita, from the Moon, on Earth!

---

1. "Mind Over Your Matter," Feb 8, 2014, http://larouchepac.com/node/29786

Made in the USA
Columbia, SC
21 November 2020